AF333560

Progress in Molecular and Subcellular Biology

Series Editors: Ph. Jeanteur, I. Kostovic, Y. Kuchino,
W.E.G. Müller (Managing Editor), A. Macieira-Coelho,
R.E. Rhoads

21

Springer

Berlin
Heidelberg
New York
Barcelona
Budapest
Hong Kong
London
Milan
Paris
Singapore
Tokyo

Werner E.G. Müller (Ed.)

Molecular Evolution: Towards the Origin of Metazoa

With 39 Figures

Springer

Professor Dr. WERNER E.G. MÜLLER

Institut für Physiologische Chemie
Abteilung für Angewandte Molekularbiologie
Johannes Gutenberg-Universität
Duesbergweg 6
D-55099 Mainz
Germany

ISBN 3-540-64565-9 Springer-Verlag Berlin Heidelberg New York

Library of Congress Cataloging-in-Publication Data.
Molecular evolution: towards the origin of Metazoa / Werner E.G. Müller (ed.).
     p.   cm.—(Progress in molecular and subcellular biology; 21)
   Includes bibliographical references and index.
   ISBN 3-540-64565-9 (hardcover)
   1. Molecular evolution.   I. Müller, Werner E.G.   II. Series.
   QH506.P76   no. 21
   [QH325]
   572.8 s–dc21
   [572.8'38]

Cover design: Meta Design, Berlin
Typesetting: Best-set Typesetter Ltd., Hong Kong
SPIN 10569593 39/3137 5 4 3 2 1 0 – Printed on acid-free paper

Preface

The base of metazoan radiation remained unrevealed for a long time because the relationship between one major early phylum, the Porifera (sponges), without doubt simple multicellular animals, and the other Metazoa was not established unequivocally. Morphological, embryological, cytological and physiological criteria could not alone answer the question of whether or not sponges have basic structural elements in common with other animals (e.g. molecules of the basal lamina or of nerve cells). It was hoped and expected that a molecular biological approach could contribute to or even solve the problem of whether the Metazoa are of mono- or polyphyletic origin.

In a first approach, sequence data from ribosomal RNA only suggested but did not prove (Field et al. 1988; Wainright et al. 1993; Kumar and Rzhetsky 1996) that sponges are at the base of the Metazoa. Hence, in 1996 R.A. Raff concluded that "the sponges have a multicellular grade of organization, with no discrete tissues or organs" and "it has never been certain whether they are really metazoans or possibly 'parazoans', with an ancestry within the protists separate from that of true metazoans". Another view proposed (Morris 1993), or took as a fact (Ax 1995), that sponges have developed extracellular molecules charateristic of the higher Metazoa, and hence should be included into the metazoan phylogenetic tree.

For a proof it is a prerequisite to analyze, on a molecular level, the extracellular materials particularly in sponges and to compare their primary structures with those present in higher metazoan phyla. The method of choice is to isolate the nucleotide sequences coding for them and to compare their deduced amino acid sequences. This second approach of a molecular phylogeny, the analysis of amino acid sequences from protein-coding genes, was successful and led to the identification of adhesive glycoproteins (Pfeifer et al. 1993), fibronectin (Müller 1998b) and integrin (Pancer et al. 1997) in sponges. The conclusion from these data, that Porifera have basic structural and regulatory molecules in common with higher metazoan phyla (Müller 1998a,b), provided evidence for a monophyly of Metazoa.

Another important question concerning the evolution to Metazoa, the origin of cell colonies, remained unanswered. It is a fact that multicellularity has arisen several times in evolution, even among prokaryotes. With respect to Metazoa, Porifera again created problems. The class of Hexactinellida is provided with syncytial tissue, while the two other sponge classes, the Calcarea and Demospongiae, are composed of uninuclear cells only. Hence multicellularity in Porifera might have been achieved early in evolution by aggregation of mitotically (un)related, mononuclear cells and the multinucleate state is a derived state, or vice versa. A first experimental approach, again using amino acid sequences from protein-coding genes (in this context, protein kinases) from Hexactinellida provides evidence that the syncytial stage was first.

Based on experimental data, this volume tries to contribute new views – like those recently published in *Progress in Molecular Subcellular Biology* (Müller 1998c) to understand one of the most "enigmatic of all phylogenetic problems" (Willmer 1994), the origin of Metazoa.

References

Ax P (1995) Das System der Metazoa. Gustav Fischer Verlag, Stuttgart

Field KG, Olsen GJ, Lane DJ, Giovannoni SJ, Ghiselin MT, Raff EC, Pace NR, Raff RA (1988) Molecular phylogeny of the animal kingdom. Science 239:748–753

Kumar S, Rzhetsky A (1996) Evolutionary relationships of eukaryotic kingdoms. J Mol Evol 42:183–193

Morris PJ (1993) The developmental role of the extracellular matrix suggests a monophyletic origin of the kingdom Animalia. Evolution 47:152–165

Müller WEG (ed) (1998a) Molecular evolution: evidence for monophyly of Metazoa. Progr Mol Subcell Biol 19. Springer, Berlin Heidelberg New York

Müller WEG (1998b) Origin of Metazoa: sponges as living fossils. Naturwissenschaften 85:11–25

Müller WEG (ed) (1998c) Program in molecular subcellular biology, vol 19. Springer, Berlin Heidelberg New York

Pancer Z, Kruse M, Müller I, Müller WEG (1997) On the origin of adhesion receptors of Metazoa: cloning of the integrin α subunit cDNA from the sponge *Geodia cydonium*. Mol Biol Evol 14:391–398

Pfeifer K, Haasemann M, Gamulin V, Bretting H, Fahrenholz F, Müller WEG (1993) S-type lectins occur also in invertebrates: high conservation of the carbohydrate recognition domain in the lectin genes from the marine sponge *Geodia cydonium*. Glycobiology 3:179–184

Raff RA (1996) The shape of life. The University of Chicago Press, Chicago

Wainright PO, Hinkle G, Sogin ML, Stickel SK (1993) Monophyletic origins of the Metazoa: an evolutionary link to fungi. Science 260:340–342

Willmer P (1994) Invertebrate relationships. Cambridge University Press, Cambridge

Mainz, Germany   W.E.G. Müller
June 1998

Contents

**Origin and Phylogeny of Metazoans as Reconstructed
with rDNA Sequences**
J.W. Wägele and F. Rödding

**Sponges (Porifera) Molecular Model Systems to Study
Cellular Differentiation in Metazoa**
W.E.G. Müller, C. Wagner, C.C. Coutinho, R. Borojevic, R. Steffen,
and C. Koziol

**The Notion of the Cambrian Pananimalia Genome and a Genomic
       Difference that Separated Vertebrates from Invertebrates**
S. Ohno

Evolution of Metazoan Collagens
R. Garrone

Evolution of Early Metazoa: Phylogenetic Status of the Hexactinellida Within the Phylum of Porifera (Sponges)
W.E.G. Müller, M. Kruse, C. Koziol, J.M. Müller, and S.P. Leys

The Question of Metazoan Monophyly and the Fossil Record

S. Conway Morris[1]

1
Introduction

As metazoans, almost inevitably we have a chauvinistic interest in our origins. We must also acknowledge, however, that the acquisition of specialized tissues – notably muscles to enable macroscopic motility, and nerves to transmit information – have transformed the world. Nevertheless, matters in biology are seldom clear-cut and amongst the smaller metazoans there are distinct overlaps with some of the more complex protistans, perhaps most strikingly with the ciliates. This overlap is perhaps most familiar from the miniaturized representatives of the sandy meiofauna (e.g. Giere 1993), the curious and probably degenerate diphyletic mesozoans (e.g. Katayama et al. 1995; Hanelt et al. 1996; Pawlowski et al. 1996; cf Cavalier-Smith 1993, who regards mesozoans as multicellular protists), and the recently discovered *Symbion* (Funch and Kristensen 1995; Funch 1996). Moreover, the recent assignment of the endoparasitic myxozoans, long thought to be protistans (e.g. Cavalier-Smith 1993), to the Metazoa, albeit without agreement as to whether they belong within the Bilateria (Smothers et al. 1994; Schlegel et al. 1996; see also Hanelt et al. 1996 and Pawlowski et al. 1996) or Cnidaria (Siddall et al. 1995) is a timely reminder that the concept of this Kingdom is more protean than is popularly imagined.

Despite the divergence of forms, a consensus emerging in recent years has very much pointed to a monophyletic origin of the metazoans (von Salvini-Plawen 1978; Ax 1989; Schram 1991; Backeljau et al. 1993; Conway Morris 1993a; Müller 1995; Müller et al. 1995; Nielsen 1995). Schram (1991, p. 39), for example, gave as key characteristics of the metazoans "the presence of collagen [but see below], acetycholine/cholinesterase systems, 9 + 2 flagellated or ciliated sperm with condensed chromatin and mitochondria, and location of reproductive cells or tissues internally". The argument for metazoan monophyly has been reinforced in more recent years by a rather remarkable range of molecular data from a variety of primitive animals (Müller et al. 1995). In this

[1]Department of Earth Sciences, University of Cambridge, Downing Street, Cambridge CB2 3EQ, UK

Progress in Molecular and Subcellular Biology, Vol. 21
W.E.G. Müller (Ed.)
© Springer-Verlag Berlin Heidelberg 1998

latter respect particular insights have come from the following: the *ets* multigene family (Degnan et al. 1993), *ras* related genes (Bosch et al. 1995), histone gene clusters (Miller et al. 1993), membrane receptors in the form of tyrosine kinase domains (Schartl and Barnekow 1982; Ottilie et al. 1992; Gamulin et al. 1994, 1997; Schäcke et al. 1994b,c; Müller 1995) and the immunoglobin (Ig) superfamily (Gamulin et al. 1994; Schäcke et al. 1994a; Müller 1995), S-type lectins (Gamulin et al. 1994; Müller 1995), extracellular matrices (ECM) (Morris 1993; Müller 1995; see also Cassaro and Dietrich 1977; Sarras et al. 1994; Reber-Müller et al. 1995); antistasin (Holstein et al. 1992), ubiquitin (Müller 1995; Müller et al. 1994, 1995), integrins (Pancer et al. 1997), and perhaps most famously the *Hox* and other homeotic genes whose recognition in the primitive cnidarians (e.g. Schierwater et al. 1991; Murtha et al. 1991; Schummer et al. 1992; Miles and Miller 1992; Miller and Miles 1993; Shenk et al. 1993a,b; Naito et al. 1993; Aerne et al. 1995; Kuhn et al. 1996; see also Grens et al. 1995) has now been complemented by their discovery in the yet more primitive sponges (e.g. Coutinho et al. 1994; Gamulin et al. 1994; Kruse et al. 1994; Seimiya et al. 1994; Degnan et al. 1995). To this battery of molecular information should be added the specifics of the developmental processes unique to metazoans (Erwin 1993). Such information leads to a depiction of the molecular ur-metazoan, a sort of metazoan "Eve" (Shenk and Steele 1993). It is worth emphasizing that while the search for such an "Eve" is well worthwhile, a substantial proportion of existing molecular data comes from the higher animals, and the crucial observations on the sponges and placozoans, and to a lesser extent the cnidarians, are in comparatively short supply.

Nevertheless, are matters of metazoan monophyly quite as clear-cut as the imposing list of evidence given above (see also Adoutte and Philippe 1993; Morris 1993) might suggest? The molecules of haemoglobin and collagen both provide a case in point. The former molecule may be less instructive in as much as evidence grows for it being primitive, evolving long before the appearance of metazoans (e.g. Hardison 1996). The notably sporadic distribution of haemoglobin in the metazoans presumably represents widespread loss, and for reasons that remain obscure (see Willmer 1990). The case of collagen is perhaps more interesting. It has long been known that the key amino-acids used in its construction are very widespread (e.g. Isenberg et al. 1966; Aaronson 1970; Isenberg and Lavine 1973; Stern and Stern 1992), but it had long been assumed that the protein itself was uniquely characteristic of the metazoans, with well documented occurrences in even the most primitive groups (e.g. Garrone 1978; Exposito and Garrone 1990). Recently, however, collagen has been recognized in the Fungi. The authors of this discovery (Celerin et al. 1996) are careful to point out that the fungal type forms a new class, and it conceivably arose by convergence. Nevertheless, this discovery could support the proposal of a phylogenetic connection between the Metazoa and Fungi (e.g. Baldauf and Palmer 1993; Wainwright et al. 1993; Berteaux-Lecellier et al. 1995).

2
The Rise of Metazoans

This then is the background. The purpose of this essay is threefold. Two of the points are quite specific. First, despite the near-consensus for metazoan monophyly, is there any evidence to support the alternative view of polyphyletic origins? The second question to be pursued is what light, if any, might the fossil record throw on the origin(s) of the Metazoa? The final question is considerably wider, surprisingly neglected, and remarkably intractable. In brief, what are the constraints on life, and to what extent does convergence confound our attempts to establish a coherent phylogeny? This wider question has, of course, a direct bearing on the arguments for metazoan monophyly.

2.1
Are Metazoans Monophyletic?

As noted above, the monophyly of metazoans is now widely accepted. It is perhaps difficult to realize that until quite recently the opposite arguments, for polyphyly, commanded wide respect (e.g. Greenberg 1959; Grimstone 1959; Nursall 1962; Sleigh 1979; Anderson 1982; Inglis 1985). In addition, more recently some of the first molecular work on metazoan relationships also pointed towards a diphyletic origin (e.g. Field et al. 1988; Christen et al. 1991; see also Adoutte and Philippe 1993). In this respect the former paper by Field et al. (1988) was subject to strong criticism (see Field et al. 1989) and in the last few years the relevant molecular evidence has been construed almost entirely in a monophyletic context (e.g. Patterson 1989; Hendriks et al. 1990; Lake 1990; Winnepenninckx et al. 1992; Adoutte and Philippe 1993; Kobayashi et al. 1993, 1996; Wainwright et al. 1993; Raff et al. 1994). While I would agree that the evidence for metazoan monophyly is certainly strong, I am not sure it is yet overwhelming. There appear to be three principal reasons for adopting a sceptical stance.

The existing phylogenies are sometimes weakly supported e.g. by bootstrapping, at critical nodes (e.g. Kobayashi et al. 1996) and are also sensitive to the choice of outgroups and the addition of new data (e.g. Adoutte and Philippe 1993). Moreover, the phylogenies under present discussion are not always easy to reconcile with data obtained from other sources. New information will doubtless help to refine some proposals, but to date trees congruent to all data simply do not emerge, and there is no prospect of homoplasy becoming any less rampant. Thus, while some areas of the metazoan tree; such as the annelid-mollusc relationship (Ghiselin 1988; Conway Morris and Peel 1995; Kim et al. 1996) and the lophophorate-protostome connection (e.g. Halanych et al. 1995; Conway Morris 1995; Conway Morris et al. 1996) may be closer to a general agreement, overall it remains the case that many areas of metazoan phylogeny – including the inter-relationships of the primitive phyla (sponges, cnidarians, placozoans, ctenophores, and platyhelminthes; see below) – remain unresolved. This need neither surprise us nor depress us: we are in the business of erecting hypotheses.

The reason why metazoans are believed to be monophyletic is, of course, that they share a set of features, e.g. *ets* multigene family, *Hox* genes (see above), which appear not to occur outside this Kingdom. Searches amongst plausible outgroups, such as the ciliates and fungi, have been made and without success, but it needs to be observed that such a survey could not be described as exhaustive. The recent discovery of fungal collagen (Celerin et al. 1996) is an important reminder of the incompleteness of our knowledge. In addition, Erwin (1993) has also emphasized that although metazoan development has unique features, much of it is embedded in a protistan ancestry that includes the ciliates. Erwin's (1993, pp. 263–264) comments on the bearing of these observations to metazoan monophyly are well worth repeating. He writes "The intriguing but inescapable conclusion . . . is that if 'protists' possess most of the requirements for development, several lineages *could* [Erwin's emphasis] have independently achieved at least early grades of complex multicellular organization. Thus there may well have been other lineages with aspects of complex development earlier in the history of life and thus the apparent monophyly of the extant fungi, metaphyte and metazoan clades may be an historical artefact, rather than reflecting the appearance of unique innovations in the three clades". Erwin (1993) goes on to observe that a number of features thought to be unique to metazoan nervous systems have since transpired to be shared with various protists, and that others still regarded as synapomorphic to the metazoans may be so assigned because of our present ignorance of their existence in groups ancestral to the metazoans.

Even within the Metazoa the molecular evidence for monophyly has some interesting exceptions. For example, the mitochondrial DNA of cnidarians has a number of unique features, that led Wolstenholme (1992, p. 208) to conclude "Many of the peculiar unique features of metazoan mtDNAs seem to have arisen following divergence of the cnidarian line from the line that is ancestral to all other present-day invertebrates and vertebrates". In addition, the use of nucleotide sequences in histone gene clusters in support of metazoan monophyly (Miller et al. 1993) needs to be considered in a wider phylogenetic context. As these authors emphasize (p. 251), certain aspects of this histone structure occur substantially deeper in the Tree of Life, being found in the protistan *Volvox* whose relationship to metazoans is remote. Miller et al. (1993) stress that such similarities may result either from convergence or an otherwise unappreciated deep ancestry. Here, in a nutshell, we have the largely unexplored problem – of direct relevance to the question of metazoan monophyly – as to whether characters are either phylogenetically unreliable because they have evolved by convergence or actually are shared more widely than hitherto realized and thus cannot be treated as synapomorphic criteria for monophyly.

2.1.1
Primitive Metazoans

In this roster most authorities would certainly include the sponges, cnidarians, placozoans, and perhaps the ctenophores (see below). Notwithstanding the molecular evidence that these diploblastic taxa appear to form a monophyletic group (e.g. Christen et al. 1991; Adoutte and Philippe 1993), even a cursory examination of these groups makes it difficult to imagine intermediate forms (see also Anderson 1982). The most reasonable response to this observation is that such intermediates are extinct, and the groups above have undergone more than half a billion years of independent evolution. Certainly the placozoans (e.g. Grell and Ruthmann 1991) could be said to occupy some such intermediate position, but this is as much on the basis of having such a minimalist bodyplan that almost anything could be derived from it (see Ax 1989). Any understanding of the phylogenetic position of the placozoans, therefore, is going to rely on molecular data. Although there is a consistency of placement within the diploblasts, their precise position is still unresolved. Some authors (e.g. Wainwright et al. 1993; Philippe et al. 1994) propose that the placozoans are a sister group of the cnidarians. More recently Odorico and Miller (1997) have tentatively proposed a sister-relationship with the ctenophores. A more productive line of enquiry might be to attempt to include some of the so-called vendobionts from the Ediacaran assemblages (see below) into this roster of primitive metazoans.

The sponges are generally interpreted as monophyletic, even though the hexactinellid body-plan is distinct, notably on account of syncitial tissue (Mackie and Singla 1983; Reiswig and Mackie 1983; see also Bergquist 1985). One item, however, that was thought to represent a major difference between sponges and other metazoans, that of the reversal of the embryonic layers, is now known to be incorrect (e.g. Bergquist 1985; Misevic et al. 1990). At the moment, the relatively limited amount of molecular data (Lafay et al. 1992) hints at a polyphyletic distribution of sponges within the Metazoa, but the exact order of branching is far from resolved. It has also long been realized that there is a close connection between the sponges and the protistan choanoflagellates, which in turn may point to yet deeper relationships within the eukaryotes (e.g. Cavalier-Smith and Chao 1995). In the case of the sponges themselves, Woollacott and Pinto (1995) note that they are closer to the choanoflagellates than the eumetazoans, and this is supported by molecular evidence (Wainwright et al. 1993). In a related context one should also note that much has been made of the occurrence of the choanocytes not only in the sponges, but also in a variety of other phyla so indicating metazoan monophyly (e.g. von Salvini-Plawen 1978). Storch (1979) has, however, observed that these cells "originate from all three germ layers and serve different functions" and so concludes they may be convergent.

The difference between the sponges and cnidarians is profound, and is most notable in terms of morphogenesis and tissue development, especially the

nervous system (e.g. Grimmelikhuijzen and Westfall 1995) for which there is
no evidence in the sponges (Simpson 1984). A particular peculiarity of the
cnidarians is the epitomic cnidocysts, which may have been acquired by
symbiogenesis from protistan predecessors (Shostak and Kolluri 1995). It
has been traditional to ally the cnidarians with the ctenophores as the
Coelenterata, (e.g. Ax 1989), and recent molecular data give some support to
such a relationship, although the long branch length of the ctenophore *Beroe*
results in low bootstrap values (Kobayashi et al. 1996; see also Odorico and
Miller 1997). On the other hand, the anatomical dissimilarities between ctenophores and cnidarians are considerable, and on this basis it is possible that the
relationship between these two groups is remote. Indeed Nielsen (1995) went
so far as to place the ctenophores adjacent to the deuterostomes. In the context
of the question of metazoan monophyly it is important to draw attention
to an almost entirely neglected paper by Weill (1946) in which he describes
a planktonic organism *Ctenoctophrys chattoni* (Fig. 1). This is a remarkable
organism, with eight rows of comb-rows, the condition characteristic of all
living ctenophores, but otherwise appears to be a single cell with macronucleus
and micronucleus, and a ciliated margin. As Weill points out, this organism
has obvious similarities to the ctenophores, as well as a more general resemblance to the medusoid cnidarians. Nevertheless, it appears to be a protistan.
His discovery has received only passing mention (e.g. Fauré-Fremiet 1954,
1958; Corliss 1959), but no detailed redescription appears to be available.

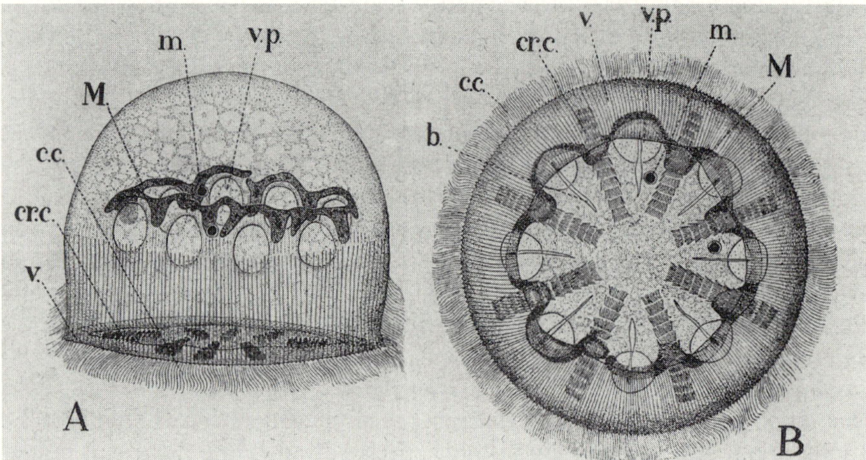

Fig. 1. The enigmatic protistan *Ctenoctophrys chattoni* from a plankton haul off Banyuls-sur-
Mer made in May 1907 by E. Chatton. **A** side view; **B** oral disc view. Abbreviations: *b* mouth
(bouche); *c.c* ciliary circlet (couronne ciliare); *M.* macronucleus (macronucléus); *m.* micronucleus (micronucléus); *v.* velum (vélum); *v.p.* pulsating vacuole (vacuole pulsatile). (Weill 1946)

Taken at face value, Weill's report would indicate the possibility of a direct protistan derivation of one metazoan phylum, the ctenophores, independent of other phyla, thereby pointing to polyphyly. Alternatively, and as Weill (p. 685) himself notes "Cette reseemblance peut résulter d'un phénomène de convergence."

Whatever the status of the sponges, cnidarians and ctenophores, there is near universal agreement that the triploblasts are monophyletic (e.g. Adoutte and Philippe 1993; Philippe et al. 1994; but see Kobayashi et al. 1996). Of the triploblasts it is the consensus that the platyhelminthes are the most primitive (e.g. Adoutte and Philippe 1993; Philippe et al. 1994). Nevertheless, the inter-relationships of the platyhelminthes and the question of their mono-phyly are by no means resolved. Recently, Haszprunar (1996) has reconsidered various data to conclude that the platyhelminthes are paraphyletic. In this scheme the most primitive group is represented by the Acoelomorpha (which is equivalent to the acoels and nemertodermatids), whereas the Catenulida are interpreted as the most advanced and probably the sister-group of all other Bilateria. Haszprunar's (1996) analysis, however, was only margin-ally concerned with molecular data (but see the Katayama et al. 1996 analysis of the 18S rDNA data that support a primitive position for the acoels; see also Hanelt et al. 1996). At first sight it may be significant that very recently some doubt has been cast on the thesis of platyhelminthes being primitive, i.e. a sister group to the Bilateria, by the molecular work (18S RNA and *Hox* genes) of Balavoine (1997), who argues for their having a position closer to the protostomes. Nevertheless, Balavoine (1997, p. 91) was careful to point out that "The protostomian emergence [of the platyhelminths] cannot be extended to the acoel flatworms, since it is not established by the 18S [RNA] data that flatworms in the broader sense are actually monophyletic". Such views also appear to be congruent with those expressed earlier by Rohde et al. (1988) who queried whether the Acoelomorpha and the enigmatic and possibly more primitive *Xenoturbella* genuinely belonged to the platyhelminth clade. Further work, therefore, may raise interesting questions concerning the nature of the most primitive triploblastic metazoans, the position of the platyhelminthes, and would also help to constrain discussion of their deriva-tion from more primitive stocks. In this sense, the analysis by Haszprunar (1996) is important because it attempts to provide some new insights into the transition between the diploblasts and triploblasts. Thus, his proposal that the acoelomorphs represent the most primitive platyhelminthes (an idea by no means widely shared e.g. Smith and Tyler 1985) may be congruent with certain species within this group possessing "a simple neural plexus like those in Cnidaria and Ctenophora", although others do exhibit some sort of brain. Other features that may be very primitive are the absence of protonephridia and very simple gonads. Nevertheless, while these characters are intriguing, there still remains a substantial gulf between such organisms and the diploblasts.

3
Fossil Evidence for the Early Evolution of Metazoans

The earliest indisputable fossil remains of metazoans date from the Vendian, which has an approximate time span of 600–550 million years (see Grotzinger et al. 1995 for important data constraining the younger age). The nature of the fossil assemblages, known as Ediacaran, has been the subject of debate. All are agreed, however, that the associated trace fossils are the product not only of metazoan activity, but of animals of a bilaterian grade (e.g. Bergström 1990). It is likely, however, that the diversity of Ediacaran traces (e.g. Crimes 1994) has been considerably overestimated (S. Jensen, pers. comm.). The status of the associated body fossils is, however, much more controversial. The recent recognition of Ediacaran sponges (Gehling and Rigby 1996), which was anticipated by the identification of sponge spicules (Steiner et al. 1993) and biomarkers (McCaffrey et al. 1994), in sediments of equivalent age is consistent with their primitive status. In addition, the presence of anthozoan cnidarians now seems to be well established (e.g. Conway Morris 1993b), and there is also some evidence that other fossils should be interpreted as triploblasts (e.g. Conway Morris 1993a, 1994; Waggoner 1996). There still remains a stubborn core of forms, exemplified by taxa such as *Ernietta* and *Pteridinium* (e.g. Crimes and Fedonkin 1996) whose affinities remain highly enigmatic. These remain the best candidates for the existence of a distinct group, that Seilacher (1989, 1992) has referred to as the Vendobionta. As originally formulated, neither the position of this group within the Tree of Life nor its status (?Kingdom) was specified by Seilacher. More recently there has been a shift in emphasis, with a proposal that the vendobionts are the sister group of the Eumetazoa (Buss and Seilacher 1994). Nevertheless, the interpretation of Ediacaran fossils remains intractable. In part this is because of the relative paucity of diagnostic features, and also because the reasons for preservation in sediments of a type e.g. sandstone, otherwise quite atypical for soft-part preservation are unresolved. It is these two factors that in large part led to the formulation of the influential Vendobionta hypothesis. In the context of a discussion of metazoan origins, there may be new avenues to explore. In reviewing the topic of metazoan development, Erwin (1993) has already noted the possibility that the Ediacaran fossils represent a separate multicellular offshoot, as suggested by Seilacher (1989, 1992), that built upon a developmental programme already available in the protistans. On the other hand, if it is finally agreed that these fossils represent metazoans, then these Ediacaran remains could provide crucial evidence for the debate on metazoan monophyly. For example, if evidence for polyphyly of the metazoans does emerged, then are might ask if any of the Ediacaran fossils could be placed in the context of sponge, cnidarians, ctenophore or triploblast origination? Alternatively, if the hypothesis of metazoan monophyly continues to hold the day, then could any of the Ediacaran fossils be informative about the intermediates between these groups? Finally, could some of the Ediacaran taxa represent intermediates between the Metazoa and their presently identified sister-groups, notably the fungi and/or the choanoflagellates?

4
The Search of Pre-Ediacaran Metazoans

The oldest recognized Ediacaran fossils appear in inter-tillite (glacial) beds in north-west Canada (Hofmann et al. 1990). The great majority, however, occur in the post-tillite faunas, and indeed the most diverse assemblages seem to have flourished shortly before the Cambrian. It has long been recognized that the earliest metazoan fossil records must be preceded by some sort of prior history, but it has remained uncertain whether such an interval was, in terms of geological time, brief or protracted (e.g. Durham 1978; Conway Morris 1993a; Fortey et al. 1996). Certainly most of the claims for pre-Ediacaran metazoans, principally in the form of trace fossils or medusoids, have failed to withstand scrutiny. New data arising from molecular biology are providing valuable insights that point to an origination at least 750 million years ago (Runnegar 1982, 1985). Such estimates are based on molecular clocks, and more recently Wray et al. (1996) have employed this methodology on a wider variety of proteins and 18S RNA, concluding that the diversification of metazoans occurred over 1000 million years ago. A critical assessment of their data (Conway Morris 1997) indicates that while a pre-Ediacaran origination is supported, and in itself is hardly surprising, it may not be nearly as deep as originally concluded. In particular, the enormous latitude in the estimated dates of divergence using different clocks suggests that simple averaging, as employed by Wray et al. (1996) is too imprecise a method. The data based on some of the proteins, notably β haemoglobin, NADH 1 and 18S RNA, demonstrates that their clocks run consistently faster (for reasons that at present are obscure) and may seriously over-estimate divergence times. Cytochrome oxidase I and ATPase 6, however, give figures that, when applied to the vertebrates with a good fossil record, are congruent with the stratigraphic divergence times. Applied to the question of the divergence of the metazoans, these slower clocks point to figures of divergence in the order of 750 million years ago, a figure that is in agreement with a parallel study of molecular clocks that was published slightly earlier by Doolittle et al. (1996).

There remains the scandal of how palaeontologists have overlooked this cryptic interval of metazoan evolution. There are, however, some constraints. The absence of large trace fossils, such as burrows, is compelling evidence that whatever metazoans were present were small, and lacked macroscopic musculature and the associated neurology to employ burrowing or walking cycles. It may also be that our search images for primitive metazoans have been misapplied. Although the microfossil record of the Neoproterozoic is now quite well documented, there are both remains of possible skeletal plates (Allison and Hilgert 1986) and enigmatic fossils such as *Valkyria* (Butterfield et al. 1994) that may require reassessment in the light of early metazoan evolution. To date, however, speculation on pre-Ediacaran metazoans has been placed in a strongly uniformitarian framework, with emphasis on comparisons with either the meiofauna or planktonic larvae.

5
Where Do We Go from Here?

Several avenues to address further the question of metazoan monophyly are evidently worth exploring.

5.1
What is the Sister Group (or Sister Groups) of the Metazoans, and Will the Fossil Record Yield any Insights?

In terms of the closest relatives of the Metazoa, the evidence continues to point towards the Fungi (e.g. Baldauf and Palmer 1993; Wainwright et al. 1993; von Ossowski et al. 1993; Nikoh et al. 1994; Doolittle et al. 1996; Kumar and Rzhetsky 1996; see also Raff et al. 1994). The other widely espoused idea, that of a metazoan-algae/plant relationship (e.g. Aaronson 1970; Isenberg and Lavine 1973; Gouy and Li 1989) is accordingly less likely (but see Gupta 1995 for a critique of the conclusions thus far drawn). In passing, one might also note that Inglis (1985), in arguing against metazoan monophyly, has gone so far as to suggest that animals represent an evolutionary grade with one branch deriving from relatives of the fungi and the other from the plants. Accepting the former proposal of a fungal relationship, the nature of this common ancestor appears not to have been explored in any depth. The report by Cavalier-Smith and Chao (1995) in this context is intriguing, involving as it does the identification of an opalozoan (*Apusomonas*) as related to the common ancestor of not only the fungi and metazoans, but also the choanoflagellates which are, of course, widely interpreted as the sister group of the sponges. That our concept of the choanoflagellates is also far from complete is apparent from the report by Cavalier-Smith and Allsopp (1996). These workers reassess a hitherto "enigmatic non-flagellate and non-photosynthetic protist" known as *Corallochytrium limacisporum* and conclude, on the basis of its 18S RNA gene, that it must be closely related to the choanoflagellates, as well as the intracellular parasite of the salmon known as the rosette agent (see Kerk et al. 1995). Cavalier-Smith and Allsopp (1996) argue that *Corallochytrium* is degenerate, losing its cilia, and coming to mimic the unicellular fungi.

5.2
What Are the Inter-Relationships of the "Primitive" Metazoans, Notably the Sponges, Cnidarians, and Perhaps the Ctenophores?

As noted above, the phylogeny of the basal part of the metazoan tree and the nature of the intermediate taxa that linked these now disparate groups is controversial. Thus, although presenting data in support of metazoan monophyly, Morris (1993) still emphasized the profound differences between the three great branches of sponges, cnidarians and the triploblasts. Each effectively pursues a separate cellular strategy of respectively "mesenchymal cell specialists" (sponges), those with a "bewildering diversity of epithelial cell types" (cnidarians), and the third branch "which have grouped cells into tissue and organ systems". Thus, as I emphasized earlier (Conway Morris 1993a) the undoubted importance of molecular data should not overshadow the fact that the anatomical, functional and ecological descriptions of the intermediate forms remain for the large part unsolved and, if extinct, only available through the fossil record. Here, it is possible that the fossil record will be of peculiar significance, and certain so-called vendobionts may play a key role.

5.3
Metazoan Evolution and Convergence: What Are the Constraints?

As in any phylogenetic discussion, and especially a cladistic formulation, the identification of monophyly presupposes the identification of reliable characters and the refutation of homoplasy. Widely acknowledged, but too often unremarked, are the problems of evolutionary convergence, the ubiquity of which suggests that the constraints on biological expression may be much more severe than generally appreciated. In the context of metazoan evolution and monophyly, one can turn to the prescient remarks of Grimstone (1959) who observed, in the context of metazoan polyphyly, that the restricting number of possible solutions to a particular problem may lead to rampant convergence. It is also interesting to note than in Morris' (1993) thoughtful review, while supporting metazoan monophyly, he is also careful to consider the likelihoods of convergence. This topic was also explored in a paper by Morris and Cobabe (1991) where they presented a critique on evolutionary convergences, emphasizing the likelihood of polyphyletic derivation of identical molecules but noting also that the biosynthetic pathways to the indistinguishable end-product may be very different. Indeed, as convergence is clearly rampant *within* the metazoans (e.g. Wake 1991; Hall 1996; Moore and Willmer 1997), one is entitled to ask whether it can be dismissed for the metazoans as a whole. In the introduction to this paper, the long list of molecular evidence cited in support of metazoan monophyly was marshalled. The apparent absence of equivalent structures in credible sister-groups is important evidence, albeit incomplete. Moreover, in their brief discussion by way of reply to earlier criticisms Field et al. (1989, p. 55) noted that "Protists possess many of the precursors for the parallel evolution of such structures as muscle and nervous tissue. The close resemblance of these structures in cnidarians and other animals could reflect ancestry, or it could reflect the constraints of such systems regardless of origin". Somewhat similar arguments were marshalled by Adoutte and Philippe (1993) who noted that a number of features critical to metazoan organization were already inherent in the protistans. In particular they wrote (p. 24) that one "can therefore speculate that the transition from unicellularity to multicellularity has been more easy to accomplish than usually thought. In this framework, the shared characteristic of multicellular organisms can be taken to reflect either the fact that they all derive from the same common ancestor that possessed these characteristics (the monophyletic scheme) or from the fact that they derive from diverse protists that independently had the potentiality to develop these characteristics (the polyphyletic scheme with parallel evolution)." Elsewhere in this paper, Adoutte and Philippe (1993, p. 18) noted that "the question of the monophyly of Metazoa is still not settled by the molecular data and it may turn out to be very difficult to solve". Accordingly, a polyphyletic formulation of metazoan origins would seek evidence for the key items associated with multicellularity, e.g. cell adhesion molecules, evolving independently from plausible protistan ancestors. It is also well to remember that on a wider scale the evidence for convergence at various levels of molecular biology continues to accumulate, and the reader's attention is drawn to the thoughtful review by Morris and Cobabe (1991). In this paper they examine the problems associated with the construction of molecules for specific functions that arise by completely different pathways. Their emphasis, therefore, is on constraint and convergence in the context of polyphyly. There is also related information from other quarters of molecular biology. These include evidence from: histone genes (Waterborg and Robertson 1996), mutations (Cunningham et al. 1997), secondary structures of mitochondrial transfer RNA (Macey et al. 1997), proteins (Wells 1996; Govindarajan and Goldstein 1996), nucleic acid binding molecules (Graumann and Maraherl 1996), and perhaps most famously the stomach lysozymes of mammals (Stewart and Wilson 1987; Swanson et al. 1991; Messier and Stewart 1997).

Further speculation in this area is constrained by two factors. First, although open to empirical observation, at present there is a general lack of knowledge of the genetic under-pinnings of the metazoans and their close eukaryote relatives. In certain cases it may transpire that the apparently independent evolution of various characters is a misreading of derivation from a common genetic basis. Second, and more intractably, there is no complete theory to explain evolutionary convergence and the constraints of biology. The ubiquity of convergence should, however, give us pause for thought, especially in the light of molecular examples. It is possible to take a hard view on convergence and claim that in order to become a metazoan, certain aspects of morphological and anatomical architecture are essentially invariant, because only one solution to the problem of becoming a macroscopic, tissue-differentiated, heterotrophic organism is available. This is obviously a somewhat startling claim, but it is not yet possible to refute it decisively. It is worth remembering, therefore, that identity of form, whether expressed at molecular or anatomical level, is known to arise in some cases by convergence and in others by shared ancestry. Appeals to the weight of evidence are not in themselves logical, especially if evolutionary constraint is in fact very strong.

The only objective method is to record the history, which is in principle knowable and unique. For what it is worth my own belief is that metazoans are indeed monophyletic, but to my mind the argument is not yet won. More importantly, the question of biological constraint, the prevalence of convergence and the inevitability of polyphyly are not only interconnected topics, but are also unjustly neglected. In part I suggest this is because of the atomistic emphasis now given to biology, as well as an obsession with cladistic methodology which although freely acknowledging homoplasy regards it as an irritating diversion rather than a profoundly interesting problem in its own right.

Acknowledgements. I thank Sandra Last for typing numerous versions of this paper, and Dudley Simons for help with photography. Cambridge Earth Sciences Publication 5052.

References

Aaronson J (1970) Molecular evidence for evolution in the algae: a possible affinity between plant cell walls and animal skeletons. Ann N Y Acad Sci 175:531–540

Adoutte A, Philippe H (1993) The major lines of metazoan evolution: summary of traditional evidence and lessons from ribosomal RNA sequence analysis. In: Pichon Y (ed) Comparative molecular neurobiology. Birkhäuser, Basel, pp 1–30

Aerne BL, Baader CD, Schmid V (1995) Life stage and tissue-specific expression of the homeobox gene *cnoxl-pc* of the hydrozoan *Podocoryne carnea*. Dev Biol 169:547–556

Allison CW, Hilgert JW (1986) Scale microfossils from the early Cambrian of northwest Canada. J Paleontol 60:973–1015

Anderson DT (1982) Origins and relationships among the animal phyla. Proc Linn Soc NSW 106:151–166

Ax P (1989) Basic phylogenetic systemization of the Metazoa. In: Fernholm B, Bremer K, Jörnvall H (eds) The hierarchy of life. Molecules and morphology in phylogenetic analysis. Excerpta Medica, Amsterdam, pp 229–245

Backeljau T, Winnepenninckx B, de Bruyn L (1993) Cladistic analysis of metazoan relationships: a reappraisal. Cladistics 9:167–181

Balavoine G (1997) The early emergence of platyhelminthes is contradicted by the agreement between 18S rRNA and *Hox* genes data. C R Acad Sci III 320:83–94

Baldauf SL, Palmer JP (1993) Animal and fungi are each others closest relatives: congruent evidence from multiple proteins. Proc Natl Acad Sci USA 90:11558–11562

Bergquist PR (1985) Poriferan relationships. In: Conway Morris S, George JD, Gibson R, Platt HM (eds) The origins and relationships of lower invertebrates. The Systematics Association, spec vol 28. Clarendon Press, Oxford, pp 14–27

Bergström J (1990) Precambrian trace fossils and the rise of bilaterian animals. Ichnos 1:3–13

Berteaux-Lecellier V, Picard M, Thompson-Coffe C, Zickler D, Panvier-Adoutte A, Simonet J-M (1995) A nonmammalian homolog of the *PAF 1* gene (Zellweger syndrome) discovered as a gene involved in caryogamy in the fungus *Podospora anserina*. Cell 81:1043–1051

Bosch TCG, Benitez E, Gellner K, Praetzel G, Salgado LM (1995) Cloning of a *ras*-related gene from *Hydra* which responds to head-specific signals. Gene 167:191–195

Buss L, Seilacher A (1994) The phylum Vendobionta: a sister group of the Eumetazoa? Paleobiology 20:1–4

Butterfield NJ, Knoll AH, Swett K (1994) Paleobiology of the Neoproterozoic Svanbergfjellet formation, Spitsbergen. Fossils Strata 34:1–84

Cassaro CMF, Dietrich CP (1977) Distribution of sulphated mucopolysaccharides in invertebrates. J Biol Chem 252:2254–2261

Cavalier-Smith T (1993) Kingdom Protozoa and its 18 phyla. Microbiol Rev 57:953–994

Cavalier-Smith R, Allsopp MTEP (1996) *Corallochytrium*, an enigmatic non-flagellate protozoan related to choanoflagellates. Eur J Protistol 32:306–310

Cavalier-Smith R, Chao EE (1995) The opalozoan *Apusomonas* is related to the common ancestor of animals, fungi, and choanoflagellates. Proc R Soc Lond [Biol] 261:1–6

Celerin M, Ray JM, Schisler NJ, Day AW, Stetter-Stevenson WG, Laudenbach DE (1996) Fungal fimbriae are composed of collagen. EMBO J 15:4445–4453

Christen R, Ratto A, Baroin A, Perasso R, Grell KG, Adoutte A (1991) An analysis of the origin of metazoans, using comparisons of partial sequences of the 28S RNA, reveals an early emergence of triploblasts EMBO J 10:499–503

Conway Morris S (1993a) The fossil record and the early evolution of the Metazoa. Nature 361:219–225

Conway Morris S (1993b) Ediacaran-like fossils in Cambrian Burgess Shale-type faunas of North America. Palaeontology 36:593–635

Conway Morris S (1994) Why molecular biology needs palaeontology. Development (Suppl) 1994:1–13

Conway Morris S (1995) Nailing the lophophorates Nature 375:365–366

Conway Morris S (1997) Molecular clocks: defusing the Cambrian "explosion"? Curr Biol 7:R71–R74

Conway Morris S, Peel JS (1995) Articulated halkieriids from the Lower Cambrian of North Greenland and their role in early protostome evolution Philos Trans R Soc Lond [Biol] 347:305–358

Conway Morris S, Cohen BL, Gawthorp AB, Cavalier-Smith T, Winnepenninckx B (1996) Lophophorate phylogeny (Technical comment). Science 272:282

Corliss JO (1959) Comments on the systematics and phylogeny of the Protozoa. Syst Zool 8:169–190

Coutinho CC, Vissers S, Van de Vyver G (1994) Evidence of homeobox genes in the freshwater sponge *Ephydatia fluviatilis*. In: Van Soest RWM, Van Kempen TMG, Braekman JC (eds) Sponges in time and space. Biology, chemistry and palaeontology. Balkema, Rotterdam, pp 385–388

Crimes TP (1994) The period of early evolutionary failure and the dawn of evolutionary success: the record of biotic changes across the Precambrian-Cambrian boundary. In: Donovan SK (ed) The paleobiology of trace fossils. Wiley, Chichester, pp 105–133

Crimes TP, Fedonkin MA (1996) Biotic changes in platform communities across the Precambrian Phanerozoic boundary. Riv Ital Paleontol Stratigr 102:317–332

Cunningham CW, Jeng K, Husti J, Badgett M, Molineux IJ, Hillis DM, Bull JJ (1997) Parallel molecular evolution of deletions and nonsense mutations in bacteriophage T7. Mol Biol Evol 14:113–116

Degnan BM, Degnan SM, Naganuma T, Morse DE (1993) The *ets* multigene family is conserved throughout the Metazoa. Nucleic Acids Res 21:3479–3484

Degnan BM, Degnan SM, Giusti A, Morse DE (1995) A *hox/hom* homeobox in sponges. Gene 155:175–177

Doolittle RF, Feng D-F, Tsang S, Cho G, Little E (1996) Determining divergence times of the major kingdoms of living organisms with a protein clock. Science 271:470–477

Durham JW (1978) The probable metazoan biota of the Precambrian as indicated by the subsequent record. Annu Rev Earth Planet Sci 6:21–42

Erwin DH (1993) The origin of metazoan development: a palaeobiological perspective. Biol J Linn Soc 50:255–274

Exposito JY, Garrone R (1990) Characterization of a fibrillar collagen gene in sponges reveals the early evolutionary appearance of two collagen gene families. Proc Natl Acad Sci USA 87:6670–6673

Fauré-Fremiet E (1954) Les problèmes de la différentiation chez les protistes. Bull Soc Zool Fr 79:311–329

Fauré-Fremiet E (1958) The origin of the Metazoa and the stigma of the phytoflagellates. Q J Microsc Sci 99:123–129

Field KG, Olsen GJ, Lane DJ, Giovannoni SJ, Ghiselin MT, Raff EC, Pace NR, Raff RA (1988) Molecular phylogeny of the animal kingdom. Science 239:748–753

Field KG, Olsen GJ, Giovannoni SJ, Raff EC, Pace NR, Raff RA (1989) Phylogeny and molecular data (Technical comment). Science 243:550–551

Fortey RA, Briggs DEG, Wills MA (1996) The Cambrian evolutionary "explosion": decoupling cladogenesis from morphological disparity. Biol J Linn Soc 57:13–33

Funch P (1996) The chordoid larva of *Symbion pandora* (Cycliophora) is a modified trochophore. J Morphol 230:231–262

Funch P, Kristensen RM (1995) Cycliophora is a new phylum with affinities to Entoprocta and Ectoprocta. Nature 378:711–714

Gamulin V, Rinkevich B, Schäcke H, Kruse M, Müller IM, Müller WEG (1994) Cell adhesion receptors and nuclear receptors are highly conserved from the lowest Metazoa (marine sponges) to vertebrates. Biol Chem Hoppe Seyler 375:583–588

Gamulin V, Skorokhod A, Kavsan V, Müller IM, Müller WEG (1997) Experimental indication in favor of the introns-late theory: the receptor kinase gene from the sponge *Geodia cydonium*. J Mol Evol 44:242–252

Garrone R (1978) Phylogenies of connective tissue. Morphological aspects and biosynthesis of sponge intercellular matrix. Front Matrix Biol 5:1–250

Gehling JG, Rigby JK (1996) Long expected sponges from the Neoproterozoic Ediacara fauna of South Australia. J Paleontol 70:185–195

Ghiselin MT (1988) The origin of molluscs in the light of molecular evidence. Oxf Surv Evol Biol 5:66–95

Giere O (1993) Meiobenthology. The microscopic fauna in aquatic sediments. Springer, Berlin Heidelberg New York

Gouy M, Li W-H (1989) Molecular phylogeny of the kingdoms Animalia, Plantae and Fungi. Mol Biol Evol 6:109–122

Govindarajan S, Goldstein RA (1996) Why are some protein structures so common? Proc Natl Acad Aci USA 93:3341–3345

Graumann P, Maraherl MA (1996) A case of convergent evolution of nucleic acid binding molecules. Bioessays 18:309–315

Greenberg MJ (1959) Ancestors, embryos, and symmetry. Syst Zool 8:212–221

Grell KG, Ruthmann A (1991) Placozoa. In: Harrison FW, Westfall JA (eds) Microscopic anatomy of invertebrates vol 2: Placozoa, Porifera, Cnidaria, and Ctenophora. Wiley-Liss, New York, pp 13–27

Grens A, Mason E, Marsh JL, Bode HR (1995) Evolutionary conservation of a cell fate specification gene: the *Hydra achaete-scute* homolog has proneural activity in *Drosophila*. Development 121:4027–4035

Grimmelikhuijzen CJP, Westfall JA (1995) The nervous systems of cnidarians. In: Breidbach O, Kutsch W (eds) The nervous systems of invertebrates: an evolutionary and comparative approach. Birkhäuser, Basel, pp 7–24

Grimstone AV (1959) Cytology, homology and phylogeny – a note on "organic design". Am Nat 93:273–282

Grotzinger JP, Bowring SA, Saylor BZ, Kaufman AJ (1995) Biostratigraphic and geochronologic constraints on early animal evolution. Science 270:598–604

Gupta RS (1995) Phylogenetic analysis of the 90 kDa heat shock family of protein sequences and an examination of the relationship among animals, plants, and fungi species. Mol Biol Evol 12:1063–1073

Halanych KM, Bacheller JD, Aguinaldo AMA, Liva SM, Hillis DM, Lake JA (1995) Evidence from 18S ribosomal DNA that the lophophorates are protostome animals. Science 267:1641–1643

Hall BK (1996) *Baupläne*, phylotypic stages, and constraint. Why there are so few types of animals. Evol Biol 29:215–261

Hanelt B, van Schyndel D, Adema CM, Lewis LA, Loker ES (1996) The phylogenetic position of *Rhopalura ophiocomae* (Orthonectida) based on 18S ribosomal DNA sequence analysis. Mol Biol Evol 13:1187–1191

Hardison RC (1996) A brief history of hemoglobins: plant, animal, protist, and bacteria. Proc Natl Acad Sci USA 93:5675–5679

Haszprunar G (1996) Plathelminthes and Plathelminthomorpha-paraphyletic taxa. J Zool Syst Evol Res 34:41–48

Hendriks L, van de Peer Y, van Herck M, Neefs J-M, de Watcher R (1990) The 18S ribosomal RNA sequence of the sea anemone *Anemonia sulcata* and its evolutionary position among other eukaryotes. FEBS Lett 269:445–449

Hofmann HJ, Narbonne GM, Aitken JD (1990) Ediacaran remains from intertillite beds in northwestern Canada. Geology 18:1199–1202

Holstein TW, Mala C, Kurz E, Bauer K, Greber M, David CN (1992) The primitive metazoan *Hydra* expresses antistasin, a serine protease inhibitor of vertebrate blood coagulation: cDNA cloning, cellular localisation and developmental regulation. FEBS Lett 309:288–292

Inglis WG (1985) Evolutionary waves: patterns in the origins of animal phyla. Aust J Zool 33:153–178

Isenberg HD, Lavine LS (1973) Protozoan calcification. In: Zipkin I (ed) Biological mineralization. Wiley, New York, pp 649–686

Isenberg HD, Douglas SD, Lavine LS, Spicer SS, Weissfellner H (1966) A protozoan model of hard tissue formation. Ann N Y Acad Sci 136:155–190

Katayama T, Wada H, Furuya H, Satoh N, Yamamoto M (1995) Phylogenetic position of the dicyemid Mesozoa inferred from 18S rDNA sequences. Biol Bull 189:81–90

Katayama T, Nishioka M, Yamamoto M (1996) Phylogenetic relationships among turbellarian orders inferred from 18S rDNA sequences. Zool Sci 13:747–756

Kerk D, Gee A, Standish M, Wainwright PO, Drum AS, Elston RA, Sogin ML (1995) The rosette agent of chinook salmon (*Oncorhynchus tshawytscha*) is closely related to choanoflagellates, as determined by the phylogenetic analyses of its small ribosomal subunit RNA. Mar Biol 122:187–192

Kim CB, Moon SY, Gelder SR, Kim W (1996) Phylogenetic relationships of annelids, molluscs and arthropods evidenced from molecules and morphology. J Mol Evol 43:207–215

Kobayashi M, Takahashi M, Wada H, Satoh N (1993) Molecular phylogeny inferred from sequences of small subunit ribosomal DNA, supports the monophyly of the Metazoa. Zool Sci 10:827–833

Kobayashi M, Wada H, Satoh N (1996) Early evolution of the Metazoa and phylogenetic status of diploblasts as inferred from amino acid sequence of elongation factor-1α. Mol Phylog Evol 5:414–422

Kruse M, Mikoc A, Cetkovic H, Gamulin V, Rinkevich B, Müller IM, Müller WEG (1994) Molecular evidence for the presence of a developmental gene in the lowest animals: identification of a homeobox-like gene in the marine sponge *Geodia cydonium*. Mech Ageing Dev 77:43–54

Kuhn K, Streit B, Schierwater B (1996) Homeobox genes in the cnidarian *Eleutheria dichotoma*: evolutionary implications for the origin of *Antennapedia*-class (*Hom*/Hox) genes. Mol Phylog Evol 6:30–38

Kumar S, Rzhetsky A (1996) Evolutionary relationships of eukaryotic kingdoms. J Mol Evol 42:183–193

Lafay B, Boury-Esnault N, Vacelet J, Christen R (1992) An analysis of partial 28S ribosomal RNA sequences suggests early radiations of sponges. Biosystems 28:139–151

Lake JA (1990) Origin of the Metazoa. Proc Natl Acad Sci USA 87:763–766

Macey JR, Larson A, Ananjeva NB, Papenfuss TJ (1997) Replication slippage may cause parallel evolution in the secondary structures of mitochrondrial transfer RNA. Mol Biol Evol 14:30–39

Mackie GO, Singla CL (1983) Studies on hexactinellid sponges. I. Histology of *Rhabdocalyptus dawsoni* (Lambe, 1873). Philos Trans R Soc Lond [Biol] 301:365–400

McCaffrey MA, Moldowan JM, Lipton PA, Summons RE, Peters KE, Jeganathan A, Watt DS (1994) Paleoenvironmental implications of novel C_{30} steranes in Precambrian to Cenozoic age petroleum and bitumen. Geochim Cosmochim Acta 58:529–532

Messier W, Stewart CB (1997) Episodic adaptive evolution of primate lysozymes. Nature 385:151–154

Miles A, Miller DJ (1992) Genomes of diploblastic organisms contain homeoboxes: sequences of *eveC*, an *even-skipped* homolog from the cnidarian *Acropora formosa*. Proc R Soc Lond [Biol] 248:159–161

Miller DJ, Miles A (1993) Homeobox genes and the zootype. Nature 365:215–216

Miller DJ, Harrison PL, Mahony TJ, McMillan JP, Miles A, Odorcio DM, ten Lohius MR (1993) Nucleotide sequence of the histone gene cluster in the coral *Acropora formosa* (Cnidarian; Scleractinia): features of histone gene structure and organization are common to diploblastic and triploblastic metazoans. J Mol Evol 37:245–253

Misevic GN, Schlup V, Burger MM (1990) Larval metamorphosis of *Microciona prolifera*: evidence against the reversal of layers. In: Rützler K, Macintyre W, Smith KP (eds) New perspectives in sponge biology. Smithsonian Institution Press, Washington, pp 182–187

Moore J, Willmer P (1997) Convergent evolution in invertebrates. Biol Rev 72:1–60

Morris PJ (1993) The developmental role of the extracellular matrix suggests a monophyletic origin of the Kingdom Animalia. Evolution 47:152–165

Morris PJ, Cobabe E (1991) Cuvier meets Watson and Crick: the utility of molecules as classical homologies. Biol J Linn Soc 44:307–324

Müller WEG (1995) Molecular phylogeny of Metazoa (Animals): monophyletic origin. Naturwissenschaften 82:321–329

Müller WEG, Schöder HC, Müller IM, Gamulin V (1994) Phylogenetic relationship of ubiquitin repeats in the polyubiquitin gene from the marine sponge *Geodia cydonium*. J Mol Evol 39:369–377

Müller WEG, Müller IM, Rinkevich B, Gamulin V (1995) Molecular evolution: evidence for the monophyletic origin of multicellular animals. Naturwissenschaften 82:36–38

Murtha M, Leckman J, Ruddle J (1991) Detection of homeobox genes in development and evolution. Proc Natl Acad Sci USA 88:10711–10715

Naito M, Ishiguro H, Fujisawa R, Kurosawa Y (1993) Presence of eight distinct homeobox-containing genes in cnidarians. FEBS Lett 333:271–274

Nielsen C (1995) Animal evolution. Interrelationships of the living phyla. Oxford University Press, Oxford

Nikoh N, Hayase N, Iwabe N, Kuma K-I, Miyata T (1994) Phylogenetic relationship of the Kingdoms Animalia, Plantae, and Fungi, inferred from 23 different protein sequences. Mol Biol Evol 11:762–768

Nursall JR (1962) On the origins of the major groups of animals. Evolution 16:118–123

Odorico DM, Miller DJ (1997) Internal and external relationships of the Cnidaria: implications of primary and predicted secondary structure of the 5'-end of the 23S-like rDNA. Proc R Soc Lond [Biol] 264:77–82

Ottilie S, Raulf F, Barnekow A, Hannig G, Schartl M (1992) Multiple *src*-related kinase genes, srkl + 4, in the fresh water sponge *Spongilla lacustris*. Oncogene 7:1625–1630

Pancer Z, Kruse M, Müller I, Müller WEG (1997) On the origin of metazoan adhesion receptors: cloning of integrin and subunit from the sponge *Geodia cydonium*. Mol Biol Evol 14:391–398

Patterson C (1989) Phylogenetic relations of major groups: conclusions and prospects. In: Fernholm B, Bremer K, Jörvall H (eds) The hierarchy of life. Molecules and morphology in phylogenetic analysis. Excerpta Medica, Amsterdam, pp 471–488

Pawlowski J, Montoya-Burgos J-I, Fahrni JF, Wüest J, Zaninetti L (1996) Origin of the Mesozoa inferred from 18S rRNA gene sequences. Mol Biol Evol 13:1128–1132

Philippe H, Chenuil A, Adoutte A (1994) Can the Cambrian explosion be inferred through molecular phylogeny? Development (Suppl) 1994:15–25

Raff RA, Marshall CR, Turbeville JM (1994) Using DNA sequences to unravel the Cambrian radiation of the animal phyla. Annu Rev Ecol Syst 25:351–375

Reber-Müller S, Spissinger R, Schuchert P, Spring J, Schmid V (1995) An extracellular matrix protein of jellyfish homologous to mammalian fibrillins forms different fibrils depending on the life stage of the animal. Dev Biol 169:662–672

Reiswig HM, Mackie GO (1983) Studies on hexactinellid sponges. III. The taxonomic status of Hexactinellida within the Porifera. Philos Trans R Soc Lond [Biol] 301:419–428

Rodrigo AG, Bergquist PR, Bergquist PL (1994) Inadequate support for an evolutionary link between the Metazoa and the Fungi. Syst Biol 43:578–584

Rohde K, Watson N, Cannon LRG (1988) Ultrastructure of epidermal cilia of *Pseudactinoposthia* sp (Platyhelminthes, Acoela): implications for the phylogenetic status of the Xenoturbellida and Acoelomorpha. J Submicrosc Cytol Pathol 20:759–767

Runnegar B (1982) A molecular-clock date for the origin of animal phyla. Lethaia 15:199–205

Runnegar B (1985) Collagen gene construction and evolution. J Mol Evol 22:141–149

Sarras MP, Yan L, Grens A, Zhang X, Agbas A, Huff JK, St John PL, Abrahamson DR (1994) Cloning and biological function of laminin in *Hydra vulgaris*. Dev Biol 164:312–324

Schäcke H, Müller WEG, Gamulin V, Rinkevich B (1994a) The Ig superfamily includes members from the lowest invertebrates to the highest vertebrates. Immunol Today 15:497–498

Schäcke H, Rinkevich B, Gamulin V, Müller IM, Müller WEG (1994b) Immunoglobulin-like domain is present in the extracellular part of the receptor tyrosine kinase from the marine sponge *Geodia cydonium*. J Mol Recogn 7:273–276

Schäcke H, Schröder HC, Gamulin V, Rinkevich B, Müller IM, Müller WEG (1994c) Molecular cloning of a tyrosine kinase gene from the marine sponge *Geodia cydonium*: a new member belonging to the receptor tyrosine kinase class II family. Mol Membr Biol 11:101–107

Schartl M, Barnekow A (1982) The expression in eukaryotes of the tyrosine kinase which is reactive with pp60$^{v\text{-src}}$ antibodies. Differentiation 23:109–114

Schlegel M, Lom J, Stechmann A, Bernhard D, Leipe D, Dykova I, Sogin ML (1996) Phylogenetic analysis of complete subunit ribosomal RNA coding region of *Myxidium lieberkuehni*: evidence that Myxozoa are Metazoa and related to the Bilateria. Arch Protistenkd 147:1–9

Schierwater B, Murtha M, Dick M, Ruddle FH, Buss LW (1991) Homeoboxes in cnidarians. J Exp Zool 260:413–416

Schram FR (1991) Cladistic analysis of metazoan phyla and the placement of fossil problematica. In: Simonetta AM, Conway Morris S (eds) The early evolution of Metazoa and the significance of problematic taxa. Cambridge University Press, Cambridge, pp 35–46

Schummer M, Scheurlen I, Schaller C, Galliot B (1992) Hom/Hox homeobox genes are present in hydra (*Chlorohydra viridissima*) and are differentially expressed during regeneration. EMBO J 11:1815–1823

Seilacher A (1989) Vendozoa: organismic construction in the Proterozoic biosphere. Lethaia 22:229–239

Seilacher A (1992) Vendobionta and Psammocorallia: lost constructions of Precambrian evolution. J Geol Soc Lond 149:607–613

Seimiya K, Ishiguro H, Miura K, Watanabe Y, Kurosawa Y (1994) Homeobox-containing genes in the most primitive Metazoa, the sponges. Eur J Biochem 221:219–225

Shenk MA, Steele RE (1993) A molecular snapshot of the metazoan "Eve". Trends Biochem Sci 18:459–463

Shenk MA, Bode HR, Steele RE (1993a) Expressions of *Cnox-2*, a HOM/HOX homeobox gene in hydra, is correlated with axial pattern formation. Development 117:657–667

Shenk MA, Gee L, Steele RE, Bode HR (1993b) Expression of *Cnox-2*, a HOM/HOX gene is suppressed during head formation in hydra. Dev Biol 160:108–118

Shostak S, Kolluri V (1995) Symbiogenetic orgins of cnidarian cnidocysts. Symbiosis 19:1–29

Siddall ME, Martin DS, Bridge D, Desser SS, Cone DK (1995) The demise of a phylum of protists: phylogeny of Myxozoa and other parasitic Cnidaria. J Parasitol 81:961–967

Simpson TL (1984) The cell biology of sponges. Springer, Berlin Heidelberg New York

Sleigh MA (1979) Radiation of the eukaryote Protista. In: House MR (ed) The origin of major invertebrate groups. The Systematics Association, spec vol 12. Academic Press, New York, pp 23–53

Smith J, Tyler S (1985) The acoel turbellarians: kingpins of metazoan evolution or a specialized offshoot? In: Conway Morris S, George JD, Gibson R, Platt HM (eds) The origins and relationships of lower invertebrates. The Systematics Association, spec vol 28. Clarendon Press, Oxford, pp 123–142

Smothers JF, von Dohlen CF, Smith LH, Spall RD (1994) Molecular evidence that the myxozoan protists are metazoans. Science 265:1719–1721

Steiner M, Mehl D, Reitner J, Erdtmann B-D (1993) Oldest entirely preserved sponges and other fossils from the lowermost Cambrian and a new facies reconstruction of the Yangtze platform (China). Berl Geowiss Abh E9:293–329

Stern M, Stern R (1992) A collagenous sequence in a prokaryotic hyaluronidase. Mol Biol Evol 9:1179–1180

Stewart CB, Wilson AC (1987) Sequence convergence and functional adaptation of stomach lysozymes from foregut fermenters. Cold Spring Harbor Symp Quant Biol 52:891–899

Storch V (1979) Contributions of comparative ultrastructural research to problems of invertebrate evolution. Am Zool 19:637–645

Swanson KW, Irwin PM, Wilson AC (1991) Stomach lysozyme gene of the langur monkey: tests for convergence and positive selection. J Mol Evol 33:418–425

Von Ossowski I, Hausner G, Loewen PC (1993) Molecular evolutionary analysis based on the amino acid sequence of catalase. J Mol Evol 37:71–76

Von Salvini-Plawen L (1978) On the origin and evolution of the lower Metazoa. Z Zool Syst Evol Forsch 16:40–88

Waggoner BM (1996) Phylogenetic hypotheses of the relationships of arthropods to Precambrian and Cambrian problematic taxa. Syst Biol 45:190–222

Wainwright PO, Hinkle G, Sogin ML, Stickel SK (1993) Monophyletic origins of the Metazoa: an evolutionary link with fungi. Science 260:340–342

Wake DB (1991) Homoplasy: the result of natural selection, or evidence of design limitations. Am Nat 138:543–567

Waterborg JH, Robertson AJ (1996) Common features of analagous replacement. Histone H3 genes in animals and plants. J Mol Evol 43:194–201

Weill R (1946) *Ctenoctophrys chattoni* n.g., n.sp., infusoire planctonique octoradié à caractère de méduse et de cténophore. C R Acad Sci Paris 222:683–685

Wells RS (1996) Excessive homoplasy in an evolutionarily constrainted protein. Proc R Soc Lond [Biol] 263:393–400

Willmer P (1990) Invertebrate relationships. Patterns in animal evolution. Cambridge University Press, Cambridge

Winnepenninckx B, Backeljau T, van de Peer Y, de Wachter R (1992) Structure of the small ribosomal subunit RNA of the pulmonate snail *Limicolaria kambeul* and phylogenetic analysis of the Metazoa. FEBS Lett 309:123–126

Wolstenholme DR (1992) Animal mitochondrial DNA: structure and evolution. Int Rev Cytol 141:173–216

Woollacott RM, Pinto RL (1995) Flagellar basal apparatus and its utility in phylogenetic analyses of the Porifera. J Morphol 226:247–265

Wray GA, Levinton JS, Shapiro LH (1996) Molecular evidence for deep Pre-Cambrian divergences among metazoan phyla. Science 274:568–573

The Evolution of the Lower Metazoa:
Evidence from the Phenotype

R. Rieger and S. Weyrer[1]

1
Introduction

Progress in the field of ultrastructure of the lower Metazoa over the last 3 decades, new data from RNA (16S, 18S, 28S rRNA) sequence analysis, and sequence analysis of other macromolecules such as cell-surface receptors, cell-cell- and cell-matrix- adhesion molecules or neuroactive substances, have had a profound impact on our understanding of the early steps in metazoan evolution (see Westheide and Rieger 1996; Müller et al., Chaps. 4 and 7, this Vol., for literature). As a result, characterising the nature of ancestral metazoans can now rely quite well on submicroscopical and macromolecular structures. Most important for the evaluation of metazoan phylogeny is that both structural and sequence data analysis make the monophyly of all extant metazoan taxa appear very likely (Morris 1993; Ax 1995, 1996; Müller 1995). The new data also help show the possibility of secondary derivation of certain unicellular eukaryotes from the Eumetazoa (e.g. the origin of the Myxozoa from Cnidaria; Smothers et al. 1994; Siddal et al. 1995).

As for the time frame of the early radiation of the Metazoa, palaeontological evidence suggests that unicellular eukaryotes arose from the prokaryote level about 2 billion years ago (Bengtson 1994); the origin of the multicellular Metazoa is being placed at about 0.8–1 billion years ago (Conway Morris 1993). However, a recent sequence analysis of several metazoan enzymes points to a divergence of Protostomes and Deuterostomes even earlier than that (Wray et al. 1996).

In this chapter, the term "phenotypic characters" is used for any three-dimensional feature (from anatomical to macromolecular) and/or its temporal changes in an organism. Sequence data analyses of RNAs and DNAs, in contrast, compare only linear arrangements for determining branching patterns in phylogenetic trees. The review concentrates on three questions that appear basic for the understanding of the origin of the lower Metazoa and thus of the organization of the stem species of Metazoa and Eumetazoa. The "Lower

[1]Institut für Zoologie und Limnologie, Universität Innsbruck, Technikerstraße 25, A-6020 Innsbruck, Austria

Progress in Molecular and Subcellular Biology, Vol. 21
W.E.G. Müller (Ed.)
© Springer-Verlag Berlin Heidelberg 1998

Metazoa" include the Porifera, the Placozoa and Mesozoa, as well as the "Coelenterata".

Question 1: How did the Metazoa arrive at multicellular organization?
Question 2: What lifestyles, life cycles and structural organization characterized the ancestral Metazoa?
Question 3: How did the Metazoa arrive at the diploblastic level of organization as seen in the most primitive Eumetazoa?

For each of these questions, available data, interpretations, and unresolved problems are discussed briefly.

2
Origin of Multicellular Organization

In spite of the many hypotheses trying to explain the transition from unicellu-lar organisms to the multicellular colonies of the Metazoa, only three basic models seem possible (Willmer 1990; Westheide and Rieger 1996):

1. Cell divisions of unicellular ancestors living within a common extracellu-lar matrix (ECM) of special macromolecular structure (Fig. 1A, see also below, and Müller et al., Chaps. 4 and 7, this vol.) lead to a multicellular organism.

   This evolutionary mode could have lead also to the multinucleated orga-nization of the Hexactinellida (= Symplasma), where daughter cells are not separated completely after cell divisions (Mackie and Singla 1983), but instead remain connected by specialized bridges (Westheide and Rieger 1996, p. 95). Fusion of cells must also occur in the Hexactinellida, to account for the number of cytoplasmic bridges present between the functional units (e.g. collar bodies) of the different "syncytial" networks. Although the term "syncytium" is used in the literature for this tissue, it is apparent from the original descriptions that incomplete cytokinesis (resulting in a "plasmo-dium") as well as secondary fusions (resulting in a "syncytium") cause its formation. It has yet to be established to which degree plasmodial and syncytial processes partake in the ontogenetic differentiation of the Hexactinellida.

2. Cellularization of a multinucleated cell (Fig. 1B) leads to a multicellular organism. One can envision that – in a process analogous to the superficial cleavage of most arthropod embryos – mitoses without cell divisions lead to a multinucleated cell. This plasmodium would subsequently be subdi-vided and would finally form a multicellular organism of mononucleated cells.

3. Aggregations of cells attracting each other through chemical signals (Fig. 1C) form a multicellular organism. Acrasean slime molds represent an extant model for this process. As a special case one could mention here also the symbiosis of different unicellular organisms (prokaryotes and eukary-

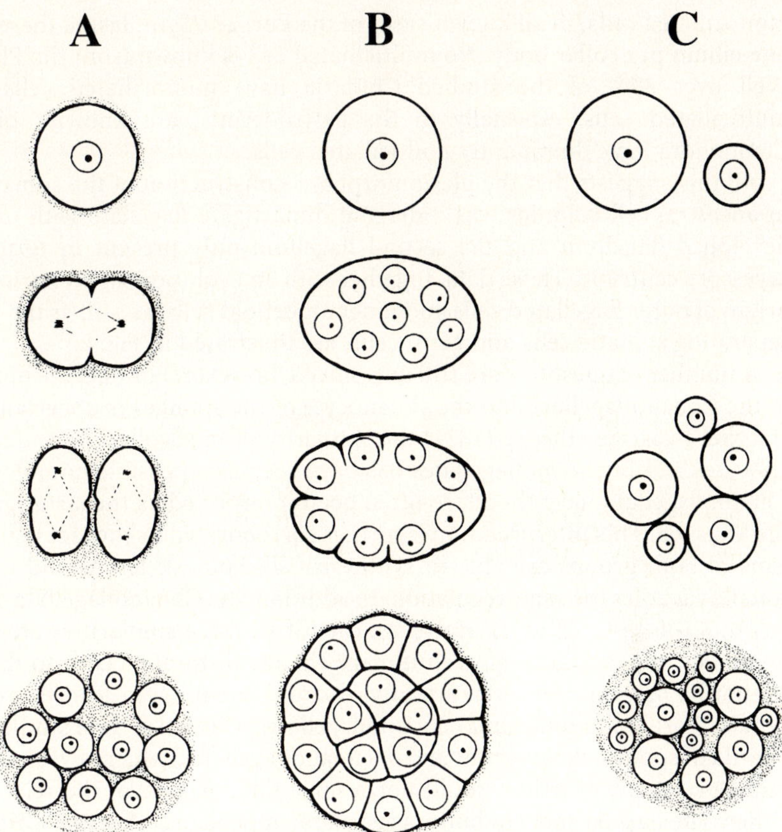

Fig. 1A–C. Processes which could have lead from unicellular eukaryotes to Metazoa. **A** Cell divisions lead to a cell colony held together by ECM (cell division colonies); **B** nuclear divisions lead to a multinucleated cell (plasmodium) which later subdivides into mononucleated cells (cellularization); **C** similar (or different) cells aggregate due to chemical stimuli and stay together because of the production of the extracellular matrix (cell aggregation). Note difference in initial occurence and location of the ECM. (from Westheide and Rieger 1996)

otes) with lower Metazoa such as the Porifera, the Placozoa and the Cnidaria.

Two sets of data, taken together, suggest that only process (1) was in effect operating during the transition from unicellular eukaryotes to the Metazoa. The first is the presence of the unique and complex macromolecular structure of the metazoan ECM already in the Parazoa and in the "Coelenterata" i.e. in the most primitive Metazoa and Eumetazoa (Har-El and Tanzer 1993; Morris 1993; Müller et al., Chaps. 4 and 7, this Vol.). The second is the pattern of occurrence of the monociliated cell type, in layers covering the organism and forming interior cavities, in the lower Metazoa. With the single exception of a sponge larva (Nielsen 1995) all known Porifera/Cellularia only have monociliated cells, in all known cases of the Porifera/Symplasma there is only one cilium per collar body. No multiciliated cell is known from the Placozoa. Well over 80% of the studied Cnidaria have monociliated cells (some multiciliated cells, especially in the gastrodermis, are known), only the Ctenophora have dominantly multiciliated cells.

All this suggests that the plesiomorphous construction of the somatic cells in ancestral cell colonies was a derived dimastigate flagellate with one fully developed flagellum and the second flagellum only present in form of an accessory centriole. These data, together with an evolutionary scenario for the origin of outer flagellated cells and inner amoeboid cells, as well as the process separating somatic cells and germ cells, are illustrated in Fig. 2.

A number of questions are still unresolved, however. For one, the homology of the Choanoflagellata and the choanocyte of the sponges is uncertain. Since Haeckel's gastraea theory (1874) the similarity of the "collar" in sponge choanocytes and in choanoflagellates has been seen as a possible homology. The Choanoflagellata have therefore often been interpreted as the sister group of the Metazoa. This interpretation is further corroborated by the fact that organisms of both groups are able to synthesize siliceous skeletons and use contractile vacuoles for osmoregulation. In addition, the Choanoflagellata do form colonies (of up to 0.5 mm). However, even if all these similarities are seen as homologs, the direction of evolutionary change from one taxon to the other cannot be determined with certainty; it could equally well lead from multicellular Porifera to unicellular choanoflagellates. Ultrastructural studies of the flagella of sponge choanocytes and of choanoflagellates, which revealed a common special substructure, could not resolve the question of the relationship either. The flagella show in both cases lateral appendages ("vanes"). However, only the vanes of choanocytes are stable enough structures to be seen in TEM-preparations while those of choanoflagellates are composed differently and are only found in shadow cast preparations. The two types of appendages are therefore presently interpreted as having evolved convergently (Ax 1995, 1996; Mehl et al. 1997).

The apparent lack of any ECM in *Trichoplax adhaerens*, the only described species of the taxon Placozoa is another problem. Considering that collagen has now been found for the first time outside of the Metazoa (in fungal fimbriae; Celerin et al. 1996), it is more than puzzling that no components to the metazoan ECM have yet been discovered in an organism which is today often seen as the most primitive in the Metazoa. In their review, Grell and Ruthmann (1991) identify the fluid surrounding the fiber cells as "probably not very different from sea water".

Finally, the secondary derivation of the "Mesozoa" from higher Eumetazoa (e.g. Platyhelminthes) remains a possibility. The lack of monociliated cells in both subtaxa of the "Mesozoa" (Dicyemida, Orthonectida), the presence of cells resembling myocytes in the Orthonectida, and the parasitic lifestyle in all species makes the interpretation that the "Mesozoa" are derived from a platyhelminth stock plausible. However, any direct evidence for this proposal is still lacking.

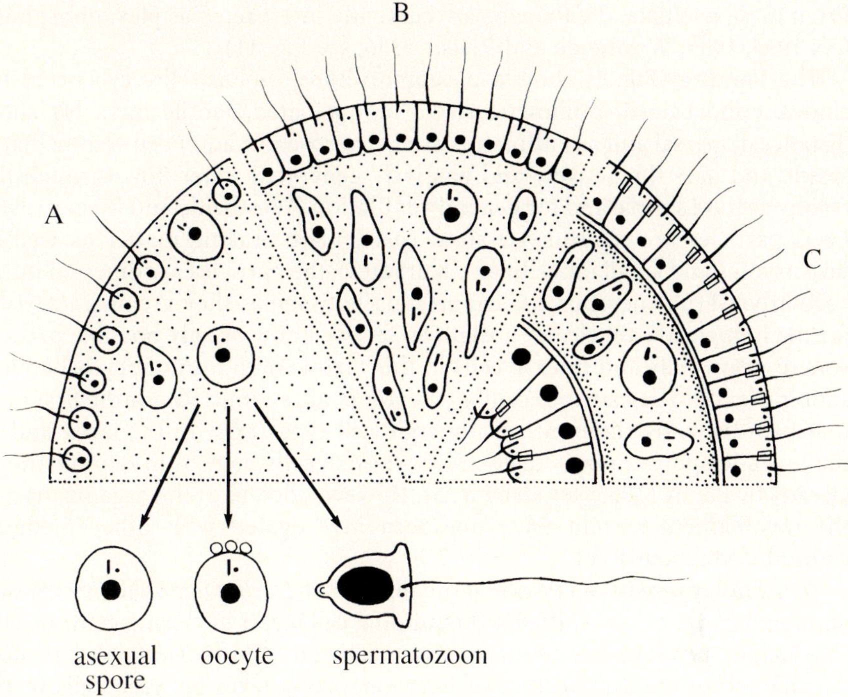

Fig. 2A–C. Hypothetical scenario of successive stages in the evolution of the early metazoan cell colonies. **A** Separation of somatic cells and three kinds of cells in the germ line in a cell division colony. Only a few somatic cells have migrated into the common ECM, they will become specialized for the secretion of the ECM components. Such cells could be envisioned as the origin of connective tissue in adults or of mesenchyme in embryos of early metazoan cell colonies. **B** Portion of a multicellular organism with incomplete "epithelium" (epithelioid layer) at the boundary of the cell colony; cells in this layer adjoin, apical junctional complexes and basal matrices are more or less missing, however. Such epithelioid layers are found in extant Porifera and Placozoa. While specializations of the ECM beneath the outer epithelioid layer (e.g. exopinacoderm) are only known within one group of Demospongiae (Pedersen 1991) molecular components of the basal lamina are already present in the ECM of Demospongia (Müller et al., Chaps. 4 and 7, this Vol.). Below the outer boundary cell layers, a connective tissue with spacious ECM is further developed when compared with **A**. **C** Eumetazoan organization of a multicellular colony, with outer epithelial layer (monociliated epidermis) and inner monociliated gastrodermis surrounding the digestive cavity, allowing the digestion of larger food items (50 μm – 1 mm) in a process of extracellular and intracellular digestion (see also Fig. 7). First signs of such a digestion are known from *Trichoplax adhaerens* (Ruthmann 1996). The central connective tissue can also be acellular (ECM without cells), as in the mesogloea of Hydrozoa, or entirely cellular (without ECM), as in certain Acoelomorpha

3
Organization, Life Cycle and Lifestyle
of the Ancestral Metazoa

One way to address this question is to consider lifestyles and life cycles in the most primitive extant Metazoa, that is in the Parazoa and in the Placozoa.

From these two taxa, the Parazoa are currently interpreted as plesiomorphous (Ax 1995, 1996; Westheide and Rieger 1996; see Fig. 11).

The Parazoa (Fig. 4) show a plesiomorphous, biphasic life cycle that includes a short-lived, millimetre-sized, monociliated, motile larva for short distance dispersal and a centimetre-sized, macroscopic adult cell colony that is sessile and uses the actively and passively generated water flow through the colony as food source and for general ventilation (Westheide and Rieger 1996). Food particles are micrometre-sized (particulate organic matter as well as prokaryotes and small unicellular eukaryotes), the particles are digested intracellularly. "True" nervous tissue and muscle tissue as they are characteristic for the Eumetazoa are clearly absent. So far, we know of only possible precursors to nerve cells and muscle cells in this taxon. With immunocytochemical staining it could be shown recently (Weyrer et al., submitted) for the first time that two distinctly different serotonergic cell types exist in the larva and in the juvenile sponge, respectively. Myocyte-like cells have been demonstrated already by Bagby (1965; see also Fig. 5). However, details of the organization of the myofilament system have not been investigated with other methods (immunocytochemistry).

In a similar way to the Porifera, the placozoan *Trichoplax adhaerens* shows an organization of two epithelioid bounding cell layers enclosing central cells. The lack of any evidence of the typical metazoan ECM in *Trichoplax* is most peculiar, since the architecture of junctional complexes between cells in the "epithelial layers" in this organism is apparently more similar to the apical junctional complex of the Eumetazoa than to junctional complexes in the epithelioid layers of sponges (Ax 1995, 1996; Westheide and Rieger 1996, p. 85). The life cycle of *Trichoplax adhaerens* is known only incompletely (Ax 1995,

---

Fig. 3A–E. Epithelial and connective tissue (mesenchymal) configuration of the monociliated somatic cell as postulated for the metazoan stem species. A1–A4 Epithelial configurations of the monociliated cell of lower Metazoa; A1 epithelioid layer of parenchymula of *Dysidea etherina*, A2 of sponge choanocyte, A3 of "epithelium" without basal matrix in *Trichoplax adhaerens*, A4 high prismatic monociliated epithelial cell of cnidarian planula larva. In Porifera, monociliated epithelioid cells of the larva apparently lack apical junctional complexes (Woollacott 1993), or have junctional complexes similar to zonulae adhaerentes (Rieger 1994b). Apical junctional complexes of the latter type occur also in *Trichoplax adhaerens* (Ruthmann 1996) and in the "Mesozoa" (Haszprunar 1996). Parazoa and Placozoa lack a typical basal matrix below their epithelioid layers. In *Trichoplax adhaerens*, none of the metazoan ECM-molecules have yet been identified. B Mesenchymal or connective tissue configuration of the monociliated cell which produces the main components of the ECM. *Arrows* indicate exchange of cells between connective tissue cells and outer epithelioid or epithelial cells. Additional connective tissue cells adjoin (*) by spot-shaped junctions (desmosomes or gap junctions). C Structure of primitive cuticles as specialization of the apical glycocalyx of a monociliated cell. D Diagram of the molecular structure of the basal matrix as appearing in vertebrates. E Diagram of macromolecular structure of the metazoan ECM and its transmembranous connection to the cell's cytoskeleton. *Ig* INTEGRIN; *aA* actin; *Ha* hyaluronic acid; *PG* proteoglycans. (Modified after Westheide and Rieger 1996)

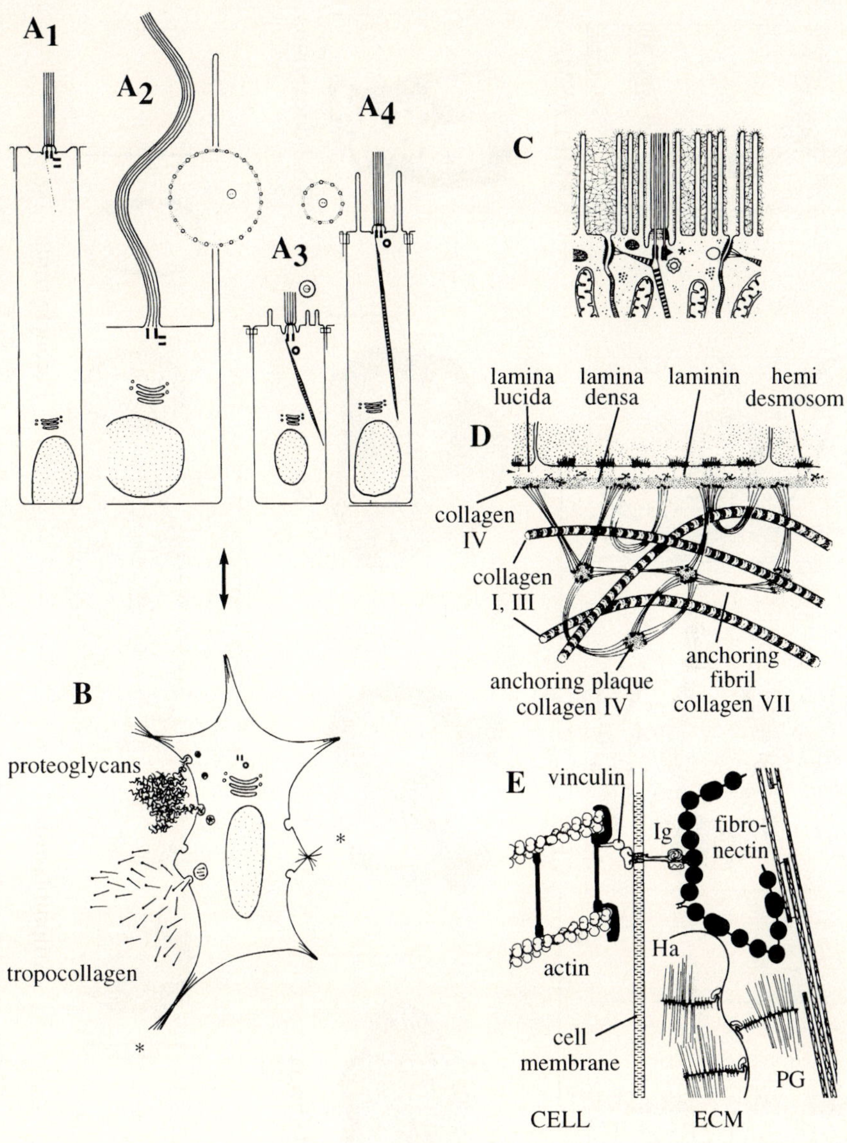
A1
A2
A3
A4
C
B
proteoglycans
tropocollagen
lamina lucida
lamina densa
laminin
hemi desmosom
D
collagen IV
collagen I, III
anchoring plaque collagen IV
anchoring fibril collagen VII
E
vinculin
Ig
fibro- nectin
actin
Ha
cell membrane
CELL
ECM
PG

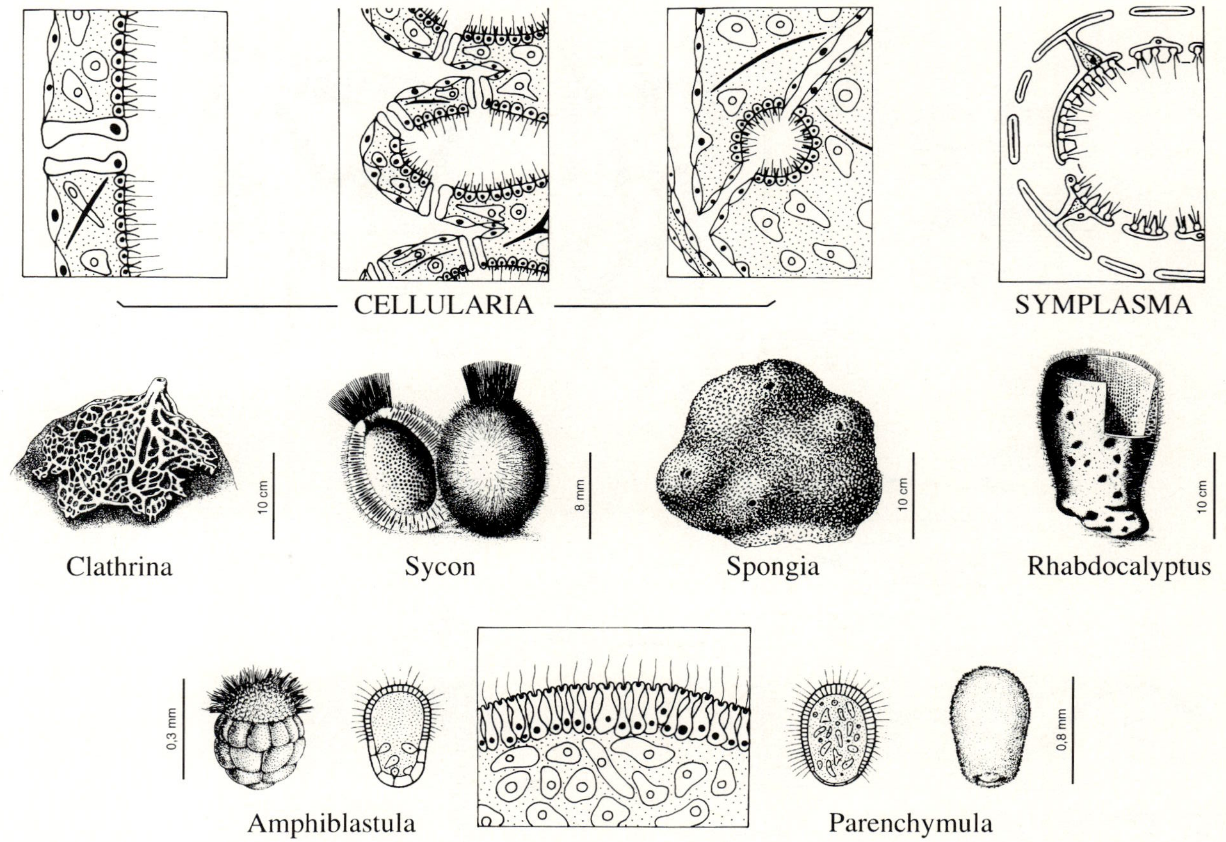

Fig. 4. Histological organization of larval and adult Porifera. *Black nuclei* identify epithelioid tissues (periderm, choanoderm). (Weyrer and Rieger; partly after van Soest 1996; Riedl 1983)

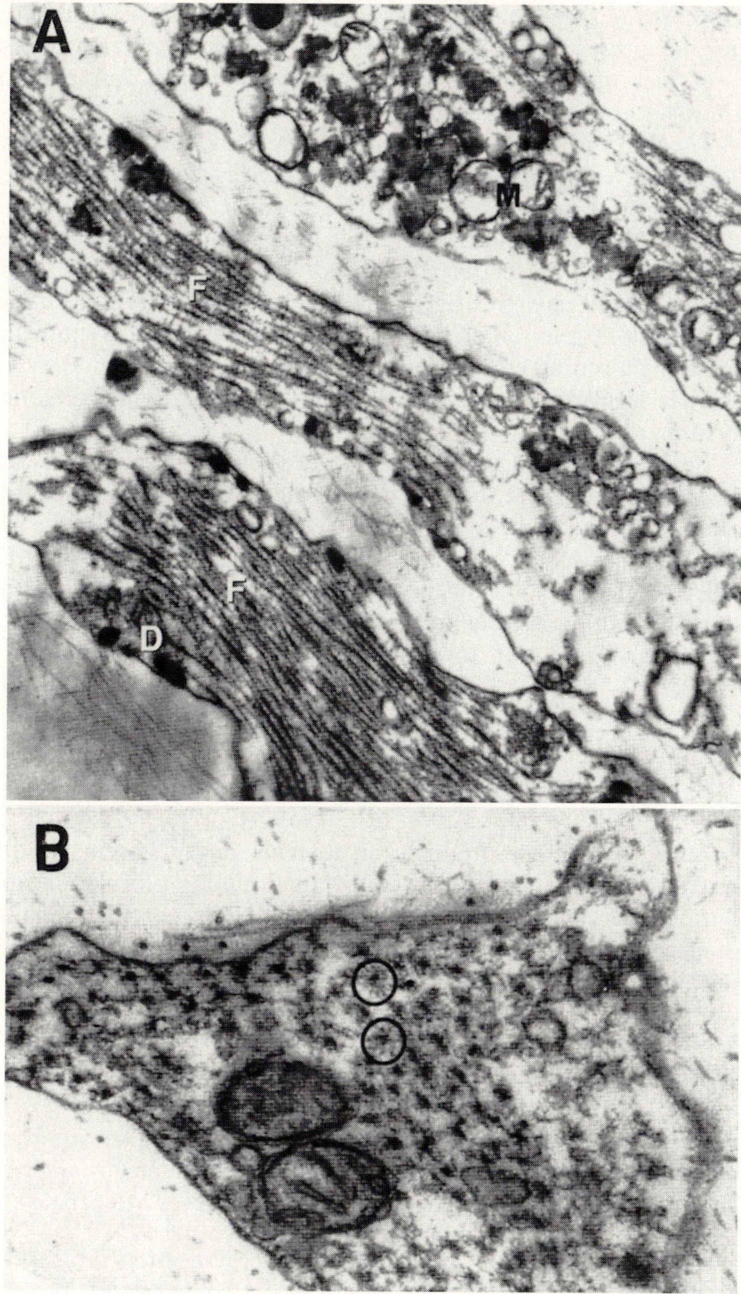

Fig. 5. **A** TEM of longitudinal section of myocyte-like cells in *Tedania ignis*; note *F* thick myofilaments, *D* dense ovoid structures (×15000) **B** TEM of cross-section through myocyte of *Microciona prolifera*, with thick and thin myofilaments (*circles*) (×80000). (from Bagby 1965)

1996). Sexual stages have not yet been shown unequivocally in natural habitats nor in cultures (Haszprunar et al. 1991). However, several modes of asexual reproduction, from simple fission to the formation of multicellular swarmers, are well known (Ruthmann 1996).

The third group of extant primitive Metazoa, the "Mesozoa", are parasitic organisms exhibiting complicated life cycles with alternating sexual and asexual modes of reproduction (Haszprunar 1996). Furthermore, the two subtaxa Dicyemida and Orthonectida most likely do not form a monophylum; the "Mesozoa" are better considered a paraphyletic – or a polyphyletic – group. A monophyletic origin cannot be established from the basically similar organization of an outer layer of at least partly ciliated epithelial cells surrounding a central cell and/or the germ cells. Even a secondary nature of the "simple" organization remains a plausible alternative interpretation (Ax 1995, 1996).

Because phylogenetic analysis of the phenotypic characters currently supports the notion that the Parazoa are the plesiomorphous sister group to all other Metazoa (Ax 1995, 1996; Westheide and Rieger 1996), the life cycle with a microscopic motile larva and a macroscopic sessile adult of the Porifera may well be seen as a model for the metazoan stem species. The observation of very slow locomotion in certain extant adult sponges could be interpreted as a relic of the early evolution towards a non-motile adult stage (Bond and Harris 1988).

4
The Origin of the Diploblastic Eumetazoa

The question of the origin of an internal cavity system for digesting and distributing nutrients and of the original mode of formation of this system during embryonic development is one of the oldest problems in the discussion of metazoan phylogeny. It is expressed in various hypotheses for the origin of the eumetazoan stem species.

4.1
Structural Innovations of the Eumetazoa

At the level of organismal construction, the following key innovations lead to eumetazoan organization and thus characterize all metazoan phyla above the level of the Parazoa, the Placozoa and possibly the "Mesozoa":

1. The emergence of "true" epithelial organization from an epithelioid layer of monociliated cells. Multicellular colonies were thereby separated from the environment more efficiently (Figs. 2, 3). The epidermis in larval cnidarians (e.g. Planula, Actinula) is a simple epithelium of tall monociliated cells. In the epidermis of adult cnidarians the cilium is usually not expressed (Harrison and Westfall 1991). Ctenophora show very

complex cells in certain areas (e.g. multiciliated comb cells). The plesiomorphous condition of the eumetazoan epithelium is thus seen most clearly in larval cnidarians.

A number of details bearing on considerations of original epithelia have been revealed by ultrastrucutral studies. Although information about characteristics of the apical junctional complex and about the macromolecular structure of the basal matrix are still lacking for many species, present data show that these epithelial characteristics are always more distinctly developed in eumetazoan organisms than in the Parazoa and the Placozoa (Ax 1995, 1996; Westheide and Rieger 1996). As is typical for many features occuring for the first time within a taxon (Riedl 1975; Wake et al. 1986), apical junctional complexes differ in the Cnidaria (a septate junction, usually without a zonula adhaerens) and the Ctenophora (a unique structure reminiscent of vertebrate tight junctions). They may represent different structural solutions for the same function: to sufficiently control the paracellular pathway (see also Willmer 1990).

2. Hand in hand with the emergence of "true" epithelial tissue goes the evolution of a central cavity for handling larger prey items in a combined process of extracellular and intracellular digestion (Fig. 6). This in turn leads to the so-called diplobastic organization of organisms, with outer epidermis and inner gastrodermis (Fig. 6). The connective tissue between the two epithelia is of epidermal origin and corresponds with the connective tissue of sponges (mesohyl). The capability to deal with new food sources through the use of special cells for adhereing to or poisoning prey (nematocysts and colloblasts) may actually have been the crucial innovation at the transition to diploblastic organization; it may have enhanced the evolution of two new cell systems, namely nervous tissue and muscle tissue.

The release of digestive enzymes into a central cavity is generally seen as the beginning of the gastrovascular system in ancestral coelenterates. Recently it has been shown, however, that in certain extant coelenterates digestion is restricted to the area of contact of prey and gastrodermis at mesenterial filaments (Bumann 1995; Bumann and Kuzirian 1996, see also Fig. 7C). It is therefore conceivable that a ventilating function was an additional important reason for the development of this organ system.

3. The development of a coordinating system in form of nerve nets, with chemical and electrical synapses (Fig. 6). Based on different neurotransmitters found, eumetazoan nervous systems are grouped in cholinergic, aminergic and peptidergic systems. In the most primitive Bilateria, all three systems are already present (Salvenmoser et al. unpublished). The "Coelenterata", however, apparently lack acetylcholine as a transmitter substance (Ali 1987). This is especially surprising because in the Porifera acetylcholin has been identified histochemically (Lentz 1968).

Possible precursors of nerve cells, that is special cells containing neuroactive substances, can also be found in the most basic metazoan organisms, below the level of the Eumetazoa. In *Trichoplax adhaerens*, special marginal

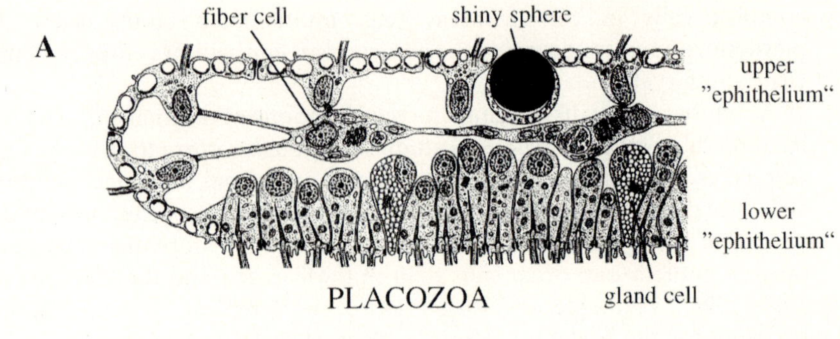

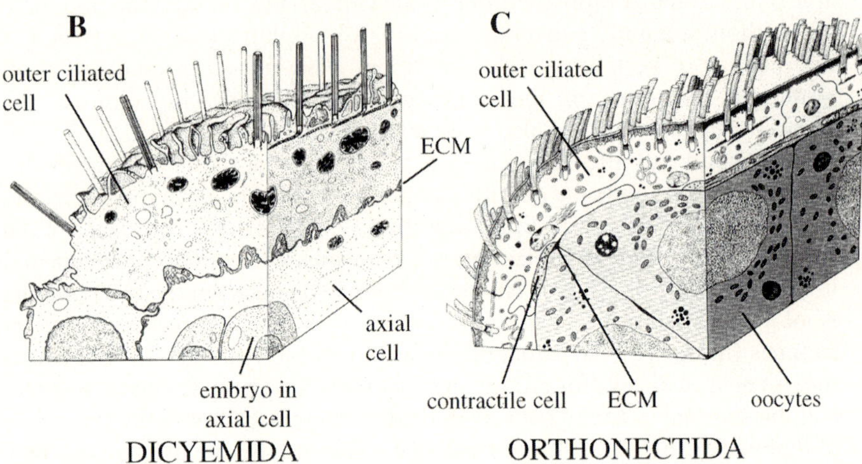

Fig. 6A–C. Histological organization of the Placozoa (**A**) and of the "Mesozoa" (**B,C**). Note especially the presence of myocyte-like cells in the "Mesozoa"-Orthonectida, the only partial presence of a basal matrix in both groups of the "Mesozoa", and the complete lack of ECM in the Placozoa. (from Westheide and Rieger 1996)

cells of the epithelioid layer are reactive to anti-FMRFamid (Schuchert 1993) and, as has been mentioned above, serotonergic cells can be demonstrated in larvae and young sponges of *Taedania ignis* (Weyrer et al., submitted).

It is generally assumed that eumetazoan nerve cells evolved from ectomdermal cells. Worthy of further studies is the possibility that the basiepithelial nerve plexus and the subepithelial nerve plexus might have evolved separately, the former as a sensoric plexus from monociliated sensory cells, the latter as a motoric plexus from special connective tissue cells within the mesoglea (Fig. 7). Nerve but cells in *Hydra* originate from stem cells (Tardent 1988) and neurons in the basal part of the epidermis may possess rudimentary cilia (Grimmelikhuijzen and Westfall 1995).

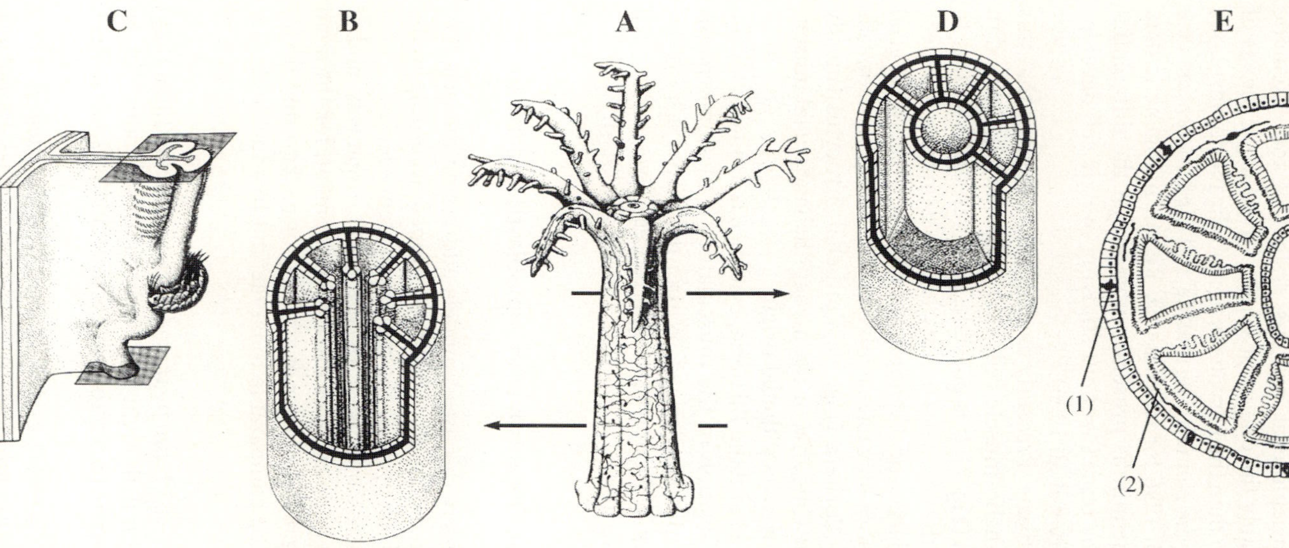

Fig. 7A–E. Diploblastic organization as seen in the plesiomorphous Anthozoa. **A** polyp in a colony of *Renilla* (Octocorallia), with primitive nerve net; **B** cut-away model of internal structure of polyp with internal septa; **C** three-dimensional diagram of mesenterial filaments, to illustrate the capture of larger food items; the most significant adaptation leading to the eumetazoan organization was the differentiation of the gastrovascular system with a single mouth/anus opening; **D** cut-away model of stomodeal region of a polyp; **E** cross-section through the polyp of *Renilla*, illustrating the location of a separate basiepithelial (1) and a subepithelial (2) nerve net. (After Ruppert and Barnes 1993; Westheide and Rieger 1996)

In the Bilateria, the basiepithelial nervous system of the enigmatic *Xenoturbella bocki* is probably the one most similar to the nerve plexus of the "Coelenterata", more so than the nervous system in the plesiomorphous free-living platyhelminths, the Nemertodermatida and the Hofsteniida (see Rieger et al. 1991; Ax 1995, 1996 for discussion). Recent ontogenetic data from the free-living platyhelminths present evidence that the differentiation of nerve cells cannot be traced without studying muscle cell development at the same time (Reiter et al. 1996).

4. The emergence of muscle tissue. Presently available data suggest that this tissue had its origin in epithelio-muscle cells that consecutively evolved to insunk fiber-type myocytes (Fig. 7). It is possible, however, that – similar to the case of the nerve nets – the two types of muscle cells stem from independent cell lines in the Cnidaria and Ctenophora: the fiber-type mesogleal muscle cells of the Ctenophora may have evolved from such precursor cells as seen already in certain Porifera (Fig. 5). Epithelial myocytes (mostly epithelio-muscle cells in the Cnidaria, or the myoepithelial parietal musculature of Ctenophora) are of myoepithelial origin.

From the Cnidaria (Fig. 8) and the Bilateria, evidence is available for the notion that the epithelial position of the musculature is plesiomorphous and that subepithelial muscles are derived. Within the Bilateria however, the fiber type muscles found in the Spiralia might represent yet another cell line that gave rise to muscle tissue (Westheide and Rieger 1996).

4.2
Models for the Most Primitive Organism with Diploblastic Organization

In the following, the most widely discussed models addressing the origin of the Eumetazoa are summerized (Fig. 9). They are grouped according to assumptions about the size and life cycle of the metazoan stem species.

Most models assume microscopic size of the adult organism and a monophasic life cycle, that is a life cycle of direct development, for the stem species of the Eumetazoa (Fig. 8; see Rieger et al. 1991 for background literature). Such organisms would be comparable to larvae, not to adults of extant eumetazoans.

1. In the Gastraea hypothesis (which can be traced back to Haeckel 1866, 1874) the diploblastic organization is thought to have been brought about by an invagination of the outer layer of a microscopic, monociliated spherical cell colony into the body.
2. The Trochea hypothesis (Nielsen and Norrevang 1985; Nielsen 1995) assumes essentially the same process to have led to the diploblastic organization.
3. In the Placula/ Bilaterogastraea hypothesis which combined Bütschli's (1884) Placula hypothesis and Jägersten's (1955, 1972) Bilaterogastraea hypothesis, the diploblastic organization is also seen to have been caused by a

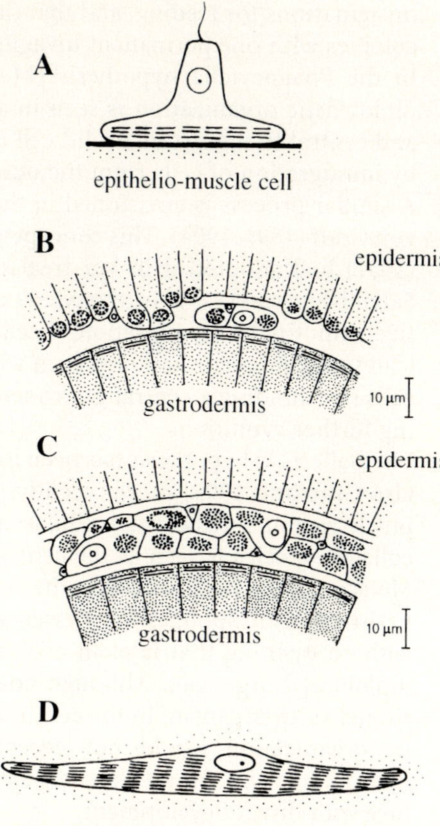

Fig. 8A–D. The origin of fiber-type subepithelial muscle cells in Cubozoa, as a model for the transition from epithelio-muscle cells to subepithelial myocytes in diploblasts. **A** Diagram with general features of epithelio-muscle cells; **B** parts of a cross section of the polyp of *Tripedalia cystophora*, with epidermal longitudinal epithelio-muscle cells and gastrodermal circular epithelio-muscle cells; it is indicated how epidermal epithelio-muscle cells sink into the mesogloea; **C** part of cross section of the polyp of *Carybdea marsupialis*, with completely submerged epidermal longitudinal fibers and gastrodermal epithelio-muscle cells; **D** generalized diagram of typical submerged fiber-type muscle cell. (Modified after Westheide and Rieger 1996)

process of invagination. It implies however, that the process occured in a disc-shaped Placula instead of a spherical Blastea. While the Placula hypothesis assumes this organism to have been of microscopic size, the Bilaterogastraea hypothesis assumes a stem species of a few centimetres, in many features resembling the extant *Xenoturbella bocki* (see Ax 1995, 1996 for literature). Grell (1971, 1981) has presented a related model on the basis of the feeding behavior of *Trichoplax adhaerens*: it implies that the lower "epithelium" of a Placula-type cell colony could have formed temporary invaginations for feeding, and that this could be seen as a transition to cell colonies with one permanent invagination.

4. In the Phagocytella hypothesis (Metschnikoff 1886), the source for the diploblastic organization is seen in a process of segregation of peripheral and central tissues from a solid cell colony. The central tissue is produced by immigration of cells from the outer cellular layer.

5. A similar process is envisioned in the Planula concept which goes back to von Graff (1891, 1904). This concept was most widely accepted after Hyman (1951) had adopted it in her treatise of the lower invertebrate taxa (see Salvini-Plawen 1978). In this view, delamination, or unipolar or alternatively multipolar immigration of cells of the outer layer is thought to have lead to a solid planuloid organism with an epidermal layer of monociliated cells surrounding a central gastrodermal tissue that became epithelial during further evolution.

6. The Gallertoid hypothesis has been introduced by Gutmann (1971, 1981; see also Grasshoff 1993) more recently. While the above concepts do not put much weight on the extracellular matrix (ECM), the ECM with fibrous collagen species is seen as the central feature in the early evolution of the Metazoa in the Gallertoid hypothesis. An internal canal system – analogous to the canals in sponges – is viewed as the starting point of a central cavity, with an opening that later in evolution forms the mouth opening of the diploblastic organism. Although not specified, the Gallertoid can be envisioned as an organism in the centimetre-range; the transition to diploblastic organization would not necessarily have occured at the level of microscopic size. However, the Gallertoid would have been a motile creature with direct development.

Contrary to these concepts, one of us has suggested (Rieger 1986, 1994a; Rieger et al. 1991) that indirect development, that is a biphasic life cycle (Fig. 10) with a motile microscopic larva and a nonmotile macroscopic adult, was characteristic for the eumetazoan stem species and that this biphasic life cycle evolved prior to the origin of tissue types (epithelioid and connective tissue, Fig. 3). Eumetazoan tissue specialization (epithelial, muscle, nervous tissue) could have therefore originated in the microscopic larva and/or in the macroscopic adult. In this model, diploblastic organization is the result of the evolution of polypoid feeding moduls from multiple invaginations (Fig. 10). Also the finding of macroscopic fossils, some of which are likely to be diploblasts, in the Ediacara Fauna (Conway Morris 1993) points to the existence of larger adult organisms already during this early phase of metazoan evolution.

5
Conclusions

The monophyletic nature of all Metazoa seems more likely since new methods have revealed several sets of characters which represent autapomorphies with a high degree of likelyhood. These characters are:

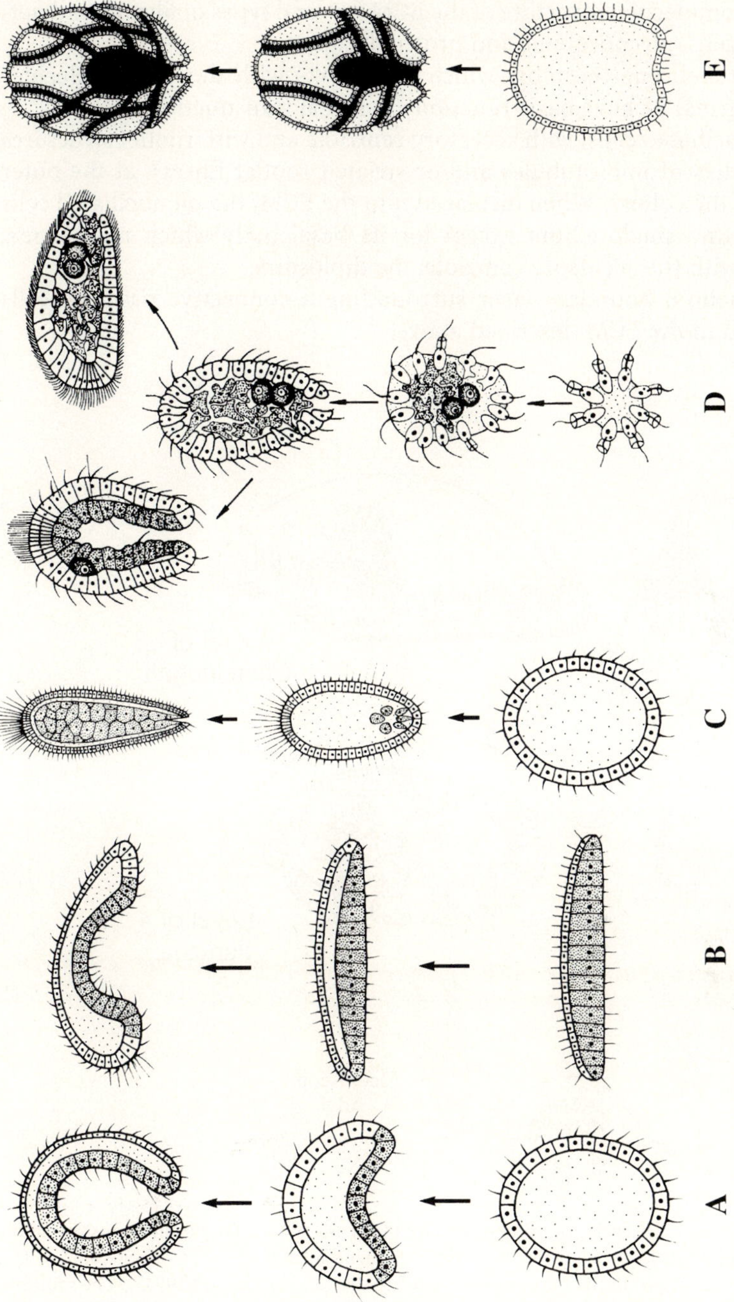

Fig. 9A–E. Models illustrating the origin of diploblastic organization. **A** Gastraea hypothesis; **B** Placula/Bilaterogastraea hypothesis; **C** Planula hypothesis; **D** Phagocytella hypothesis; **E** Gallertoid hypothesis. (After Grell et al. 1980)

1. The macromolecular structure of the ECM (several types of fibrous and net-like collagens, glycoproteins and proteoglycans).
2. Specific membrane receptors which form links with the ECM molecules (e.g. integrins) or have basic functions in signal transduction.
3. The monociliated cells, with accessory centriole and with rootlet structures (e.g. bundles of microtubules and/or striated rootlet fibres), at the outer border of the colony. When displaced into the ECM, the monociliated cells have lost the single cilium except for its basal body which now forms, together with the accessory centriole, the diplosome.
4. The epithelioid boundary layer surrounding a connective tissue of cells embedded in the ECM described above.

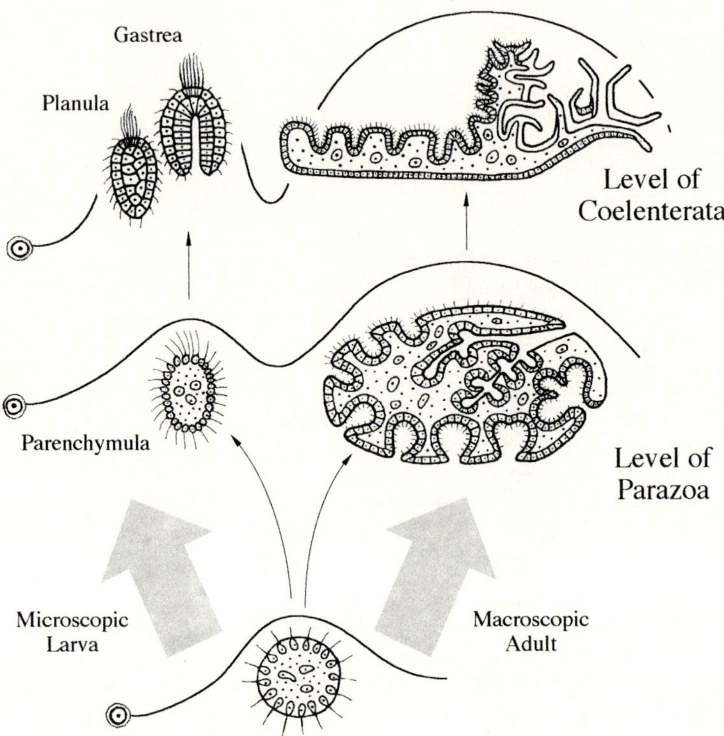

Fig. 10. The biphasic life cycle in the stem species of the Metazoa, and parallel transformation of the larval and adult body from the level of the Parazoa to that of the "Coelenterata" (other possibilities for the origin of the biphasic life cycle are discussed in Rieger 1994a). The multiple invaginations in the macroscopic adult at the level of the Parazoa developed into choanocyte chambers and a system of internal canals. In the "coelenterate" level of organization, the multiple invaginations in the adult became elaborated not only internally but also protruded at the margin of the colony where polypoid feeding modules arose. The diploblastic Gastraea or Planula is a model for the larva, not for an adult stage. (Modified after Rieger 1994a)

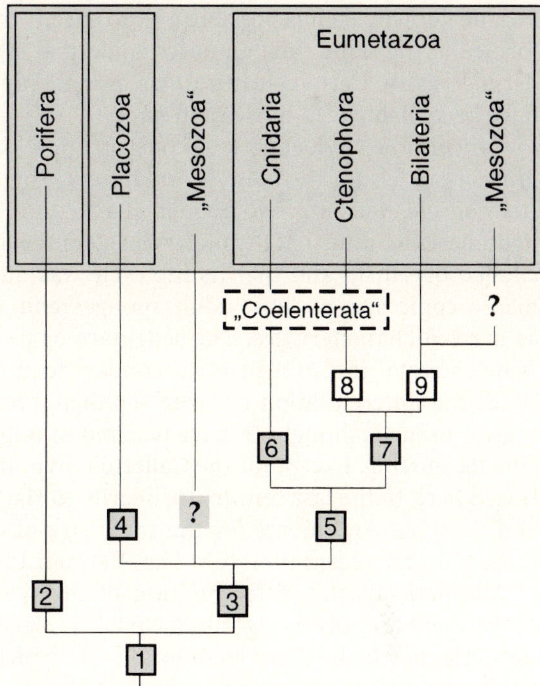

Fig. 11. Cladogram of lower Metazoa, with autapomorphies, using phenotypic characters. The two *question marks* point to the uncertain position of the "Mesozoa", either near the Placozoa or near the Bilateria. A *question mark* next to this square for apomorphies suggest that the particular apomorphies need further confirmation. 1 Meiosis at time of gamete-formation; oocyte with 3 polar bodies; spermatozoon with head-region (nucleus and acrosome), middle piece (basal body of flagellum, distal centriol, mitochondria), and tail region (axoneme of flagellum); with monociliated cells at margin of cell colony, some cells having migrated into the ECM; ECM with glycoproteins, proteoglycans and several types of collagens; with integrin-like transmembrane proteins and with signal transduction molecules; 2 Adult body covered by epithelial-like layer of pinacoderm; with various structures of internal cavities and/or globular chambers lined with choanocytes and connected to an internal system of canals that allow influx of sea water at small pores (ostia) and expulsion of sea water through larger openings (oscula); 3 apical junctional complexes at outer epithelial-like boundary layer including gland cells; 4 adults in the size-rang of millimetres, disc-shaped body covered with a monociliated "epithelium" lacking a basal matrix, lower epithelium with special gland cells used for uptake of small eukaryote cells; with star-shaped fiber cells forming a net of tissue between upper and lower "epithelium"; no ECM reported; 5 with diploblastic organization (ecto- and entodermal germ layers); with epithelial epidermis and gastrodermis, with ectomesenchyme; with central digestive cavity with mouth/ anus opening; with muscle, nerve, and sensory cells; with chemical and electrical synapses; 6 polypoid colonies with nematocysts; 7 acrosome of spermatozoon consisting of one acrosomal vesicle and a special subacrosomal structure; 8 with colloblasts and a complex apical static sensory organ; 9 with triploblastic organization (ectodermal, entodermal, entomesodermal germ layers); with formation of a brain in the anterior part of the body; with filtration kidneys (proto- and metanephridia). (After Westheide and Rieger 1996)

For the Eumetazoa it is possible now to specify that epithelial tissue apparently formed at the same time an outer epidermal layer and the lining of a gastro-dermal cavity. Certain features, such as the apical junctional complex, may have been elaborated more than once.

Concerning phylogenetic reconstructions, one can further conclude that present models for the lifestyle and life cycle of the metazoan and eumetazoan stem species fall into two groups (Bartolomaeus and Ax 1992). Traditional hypotheses assume that the metazoan stem species was a solitary, microscopic, ciliated organism and that its life cycle was one of direct development. The macroscopic size of most adult sponges and their sessile lifestyle are seen as derived characteristics. The sedentary or passively drifting lifestyle of the "Coelenterata" is also seen as a secondary feature in the primitive Eumetazoa. Thus, the differentiation of "true" epithelial tissue, of nervous tissue and of muscle tissue is thought to have occured in millimetre-sized organisms using cilia for moving. Except for the Gallertoid hypothesis, all these concepts can be traced back to the last century, primarily to Haeckel, Metschnikoff, von Graff, and Bütschli. Arguments for the small size of early metazoans are also presented in very recent literature (e.g. Vermeij 1996).

Alternatively, the differentiation of the typical histological organization of the Eumetazoa is thought to have taken place in organisms with a biphasic life cycle, in which ciliary locomotion was typical for short-lived larvae, while adult organisms were macroscopic and either sedentary or passively drifting. The adult cell colonies of the metazoan and eumetazoan stem species are thought to have originally included modules for producing water flow through the cell colony (still seen in the Porifera), that were derived from multiple invaginations and which gave rise to polypoid colonies similar to extant Octocorallia.

Molecular sequence analysis will without doubt provide new insight into the relationships among the lower Metazoa and such data will make it possible to define the evolutionary time span of the first metazoan radiations to a high degree (see Wray et al. 1996). They cannot, however, provide insights into the original lifestyle and the original lifecycle of the metazoan, specifically the eumetazoan stem species. This question remains speculative, but is central for understanding the origin of multicellular animal life.

References

Ali MA (1987) Nervous systems in invertebrates. NATO ASI, series A: Life sciences, vol 141. Springer, Belin Heidelberg New York, 675pp

Ax P (1995) Das System der Metazoa. I. Fischer, Stuttgart

Ax P (1996) Muticellular animals. A new approach to the phylogenetic order in nature. Springer, Berlin Heidelberg New York

Bagby RM (1965) The fine structure of myocytes in the sponges *Microciona prolifera* (Ellis and Solander) and *Tedania ignis* (Duchassaing and Michelotti). J Morphol 118(2):167–182

Bartolomaeus T, Ax P (1992) Protonephridia and Metanephridia – their relation within the Bilateria. Z zool Syst Evolut.-forsch. 30:21–45

Bengtson S (ed) (1994) Early life on earth. Bobel symposium 84. Columbia University Press, New York

Bond C, Harris AK (1988) Locomotion of sponges and its physical mechanism. J Exp Zool 246:271–284

Bumann D (1995) Localization of digestion activities in the sea anemone *Haliplanella luciae*. Biol Bull 189:236–237

Bumann D, Kuzirian A (1996) Direct observation of contact digestion in inside-out oriented *Aurelia aurita* polyps. Biol Bull 191:302–303

Bütschli O (1884) Bemerkungen zur Gastraeatheorie. Morphol Jahrb 9:415–427

Celerin M, Ray JM, Schisler NJ, Day AW, Stetler-Stevenson WG, Laudenbach DE (1996) Fungal fimbriae are composed of collagen. EMBO J 15(17):4445–4453

Conway Morris S (1993) The fossil record and the early evolution of the Metazoa. Nature 361: 219–225

Graff L von (1891) Die Organisation der Turbellaria Acoela. Engelmann, Leipzig

Graff L von (1904) Marine Turbellarien Otavas und der Küsten Europas. I. Einleitung Acoela. Z Wiss Zool 78:190–245

Grasshof M (1993) Die Evolution der Tiere in neuer Darstellung. Nat Mus 123(7):204–215

Grell KG (1971) *Trichoplax adhaerens* Schulze FE und die Entstehung der Metazoen. Naturwiss Rundsch 24:160–161

Grell KG (1981) *Trichoplax adhaerens* and the origin of Metazoa. Atti Conv Lincei 49:107–121

Grell KG, Ruthmann A (1991) Placozoa. In: Harrison WF, Westfall JA (eds) Microscopic anatomy of invertebrates, vol 2: Placozoa, Porifera, Cnidaria and Ctenophora. Wiley, New York, pp 13–27

Grell KG, Gruner HE, Kilian EF (1980) Einführung. In: Graner HE (ed) Lehrbuch der speziellen Zoologie, vol 1: Wirbellose Tiere. I. Einführung Protozoa, Placozoa, Porifera. The early evolution of Metazoa and the significance of problematic taxa. Fischer, Stuttgart, pp 15–16

Grimmelikhuijzen CJP, Westfall JW (1995) The nervous systems of Cnidarians. In: Breidbach O, Kutsch W (eds) The nervous systems of invertebrates: an evolutionary and comparative approach. Birkhäuser, Basel, pp 7–24

Gutmann WF (1971) Der biomechanische Gehalt der Wurmtheorie. Z Wiss Zool 182:229–262

Gutmann WF (1981) Relationships between invertebrate phyla based on functional-mechanical analysis of the hydrostatic skeleton. Am Zool 21:63–81

Haeckel E (1866) Generelle Morphologie der Organismen, vols 1, 2. Reiner, Berlin

Haeckel E (1874) Die Gastraea-Theorie, die phylogenetische Classification des Thierreichs und die Homologie der Keimblätter. Jena Z Naturwiss 8:1–55

Har-El R, Tanzer ML (1993) Extracellular matrix 3: evolution of the extracellular matrix in invertebrates. FASEB J 7:1115–1123

Harrison WF, Westfall JA (1991) Microscopic anatomy of invertebrates, vol 2: Placozoa, Porifera, Cnidaria and Ctenophora. Wiley, New York

Haszprunar G (1996) Mesozoa. In: Westheide W, Rieger R (eds) Spezielle Zoologie, part 1: Einzeller und Wirbellose Tiere. Fischer, Stuttgart, pp 125–129

Haszprunar G, Rieger RM, Schuchert P (1991) Extant "problematica" within or near the Metazoa. In: Simonetta AM, Conway-Morris S (eds) The early evolution of Metazoa and the significance of problematic taxa. Cambridge University Press, New York, pp 99–105

Hyman LH (1951) The invertebrates: Platyhelminthes and Rhynchocoela. The acoelomate Bilateria. McGraw-Hill, New York, 550pp

Jägersten G (1955) On the early phylogeny of the Metazoa. The bilaterogastraea theory. Zool Bidr Upps 30:321–354

Jägersten G (1972) Evolution of the Metazoan life cycle. Academic Press, London

Lentz TL (1968) Primitive nervous systems. Yale University Press, New Haven

Mackie GO, Singla CL (1983) Studies on hexactinellid sponges. I. Histology of *Rhabdocalyptus Dawsoni* (Lambe, 1873). Philos Trans R Soc Lond [Bioc] 301:365–400

Mehl D, Müller I, Müller WEG (1997) Molecular biological and palaeontological evidence that Eumetazoa, including Porifera (sponges), are of monophyletic origin. In: International Conference on sponge science. Springer-, Berlin Heidelberg New York (in press)

Metschnikoff E (1886) Embryologische Studien an Medusen. Ein Beitrag zur Genealogie der Primitiv-Organe. Vienna

Morris PJ (1993) The developmental role of the extracellular matrix suggests a monophyletic origin of the kingdom Animalia. Evolution 47(1):152–165

Müller WEG (1995) Molecular phylogeny of Metazoa (animals): monophyletic origin. Naturwissenschaften 82:321–329

Nielsen C (1995) Animal evolution. Oxford University Press, Oxford

Nielsen C, Norrevang A (1985) The trochaea theory: an example of life cycle phylogeny. In: Conway-Morris S, George JD, Gibson R, Platt HM (eds) The origins and relationships of lower invertebrates. Clarendon Press, Oxford, pp 28–41

Pedersen KJ (1991) Structure and composition of basement membranes and other basal matrix systems in selected invertebrates. Acta Zool 72:181–201

Reiter D, Boyer B, Ladurner P, Mair G, Salvenmoser W, Rieger R (1996) Differentiation of the body wall musculature in *Macrostomum hystericinum marinum* and *Hoploplana inquilina* (Platyhelminthes), as models for muscle development in lower Spiralia. Roux's Arch Dev Biol 205:410–423

Riedl R (1975) Die Ordnung des Lebendigen. Systembedingungen der Evolution. Parey, Hamburg

Riedl R (1983) Fauna und Flora des Mittelmeeres. Parey, Hamburg

Rieger MR (1986) Über den Ursprung der Bilateria: die Bedeutung der Ultrastrukturforschung für ein neues Verstehen der Metazoenevolution. Verh Dtsch Zool Ges 79:31–50

Rieger RM (1994a) Evolution of the "lower" Metazoa. In: Bengsten S (ed) Early life on earth. Columbia University Press, New York, pp 475–492

Rieger RM (1994b) The biphasic life cycle – a central theme of metazoan evolution. Am Zool 34:484–491

Rieger RM, Haszprunar G, Schuchert P (1991) On the origin of the Bilateria: traditional views and recent alternative concepts. In: Simonetta A, Conway Morris S (eds) The early evolution of Metazoa and the significance of problematic taxa. Cambridge University Press, Cambridge, pp 107–112

Ruppert EE, Barnes RD (1993) Invertebrate zoology, 6th edn. Saunders, Philadelphia

Ruthmann A (1996) II. Placozoa. In: Westheide W, Rieger R (eds) Spezielle Zoologie, part 1: Einzeller und Wirbellose Tiere. Fischer, Stuttgart, pp 121–124

Salvini-Plawen LV (1978) On the origin and evolution of the lower Metazoa. Z Zool Syst Evol Forsch 16:40–88

Schuchert P (1993) *Trichoplax adhaerens* (Phylum Placozoa) has cells that react with antibodies against the neuropeptide Rfamide. Acta Zool 74:115–117

Siddal ME, Martin DS, Bridge D, Desser SS, Cone DK (1995) The demise of a phylum of protists: phylogeny of Myxozoa and other parasitic cnidaria. J Parasitol 81(6):961–967

Smothers JF, von Dohlen CD, Smith LH Jr, Spall RD (1994) Molecular evidence that the myxozoan protists are metazoans. Science 265:1719–1721

Tardent P (1988) Hydra. Neujahrsblatt. Naturforsch Ges Zürich 132(5):1–100

Van Soest R (1996) Porifera, Schwämme. In: Westheide W, Rieger R (eds) Spezielle Zoologie, part 1: Einzeller and Wirbellose Tiere. Fischer, Stuttgart, pp 98–119

Vermeij GJ (1996) Animal origins. Science 274:525–526

Wake DB, Connor EF, de Ricqlès AJ, Dzik J, Fisher DC, Gould SJ, la Barbera M, Meeter DA, Mosbrugger V, Reif WE, Rieger RM, Seilacher A, Wagner GP (1986) Directions in the history of life. In: Raup DM, Jablonski D (eds) Patterns and processes in the history of life. Springer, Berlin Heidelberg New York, pp 46–67

Westheide W, Rieger R (eds) (1996) Spezielle Zoologie, part 1: Einzeller und Wirbellose Tiere. Fischer, Stuttgart, 1–909pp

Willmer P (1990) Invertebrate relationships: patterns in animal evolution. Cambridge University Press, New York

Woollacott RM (1993) Structure and swimming behavior of the larva of *Haliclona tubifera* (Porifera: Demospongiae). J Morphol 218:301–321
Wray GA, Levinton JS, Shapiro LH (1996) Molecular evidence for deep precambrian divergences among metazoan phyla. Science 274:568–573

Origin and Phylogeny of Metazoans as Reconstructed with rDNA Sequences

J.W. Wägele and F. Rödding[1]

1
Introduction

Today, students of zoology will find a large number of incompatible dendrograms of the major groups of metazoans in textbooks and articles (examples in Figs. 1 and 2) inferred from rDNA alignments. Discrepancies existed before molecular data were used: a long debate, for example, was the discussion on the monophyly of arthropods. S. Manton (1902–1979) advocated her hypotheses of polyphyly of arthropods on the basis of excellent morphological studies (e.g. Manton 1969, 1973) and it took many years of research, partly with new methods (e.g. Wheeler et al. 1993), until the majority of the scientific community accepted what was already well-founded decades before (e.g. Snodgrass 1950), namely that arthropods are with high probability monophyletic. The major source of uncertainty in comparative morphology is the lack of scientific concepts that allow a quantitative estimation of the quality (= information content) of characters and data sets. Only recently has the first author proposed that the terms "information content" and "data quality" are related to the probability of homology of the characters (Wägele 1996b).

The use of DNA sequences has the advantage that similarities can be quantified, and thus the support for alternative tree topologies is comparable with higher accuracy than when using morphological characters. The most popular sequences used for metazoan phylogeny inference are the ssu-rRNA genes, mainly for historical reasons. Of the over 4500 complete genes present in ssu-rDNA databases, about >1100 belong to eukaryotes, and >300 to metazoans (Van de Peer et al. 1996; Maidak et al. 1996). The number is increasing rapidly, but for the moment complete sequences of metazoan rRNAs are still not very numerous (only 92 metazoan 28SrRNAs in 1996: De Rijk et al. 1996).

The confidence researchers have in molecular phylogenetics based on rDNA alignments is documented by the large number of recent publications. The enthusiasm has however been dampened by the increasing number of incompatible hypotheses on relationships derived from analyses of these and other molecules (Figs. 1, 2). The present situation is not better than in the era of comparative morphology. It is obvious that *maximally one* out of several

---

[1]Fakultät für Biologie, Ruhr-Universität Bochum, 44780 Bochum, Germany

Progress in Molecular and Subcellular Biology, Vol. 21
W.E.G. Müller (Ed.)
© Springer-Verlag Berlin Heidelberg 1998

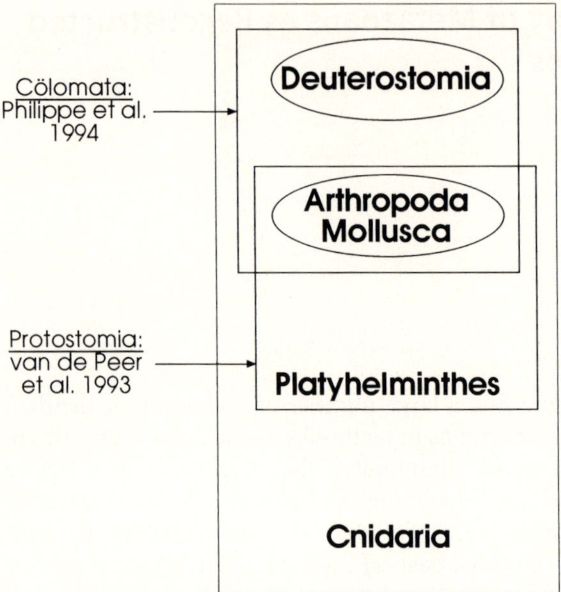

Fig. 1. Venn-diagram with a set of incompatible hypotheses derived from rDNA studies on the relationships of major metazoan groups. Philippe et al. (1994) favour a split between Cnidaria + Platyhelminthes and coelomate metazoans, while van de Peer et al. (1993) group part of the coelomates with platyhelminths

alternative tree topologies can be correct, all the other conclusions necessarily being wrong. Therefore the search for reliable methods is for the moment one of the most important tasks.

In the following, some considerations concerning sources of errors, especially the quality of data sets, and important results on metazoan phylogeny are discussed. From the published literature only some examples with studies of the phylogeny within larger metazoan groups are mentioned.

2
Data and Methods

Most information discussed in the following is found in the cited literature. With few exceptions most published dendrograms were estimated with neighbor-joining (distance) methods, maximum parsimony or maximum likelihood methods (NJ, MP or ML methods, respectively; for description of these methods see Swofford and Olsen 1990).

For the estimation of information content of ssu-rDNA (Sect. 8) we used spectra of supporting positions as described in Wägele (1996b) and Wägele and Rödding (1998). The software (PHYSID) used for signal determination

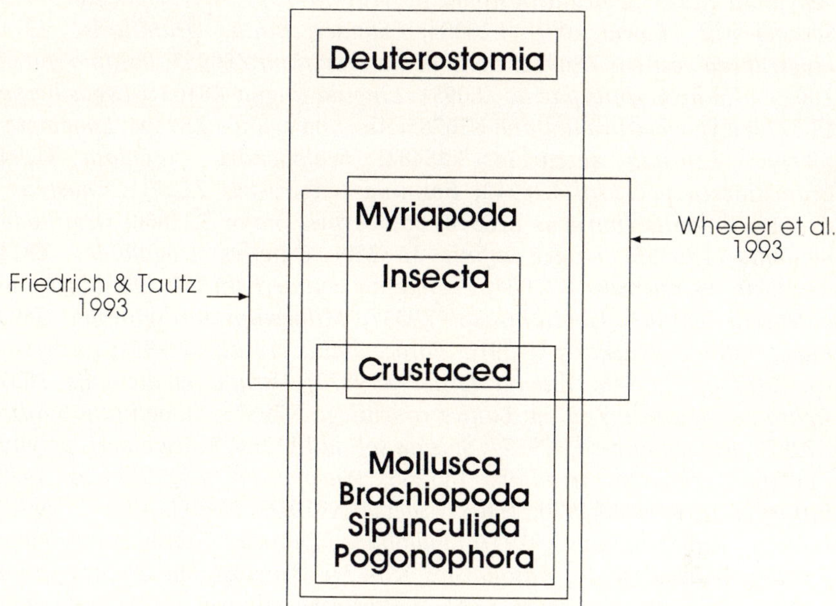

Fig. 2. Venn-diagram visualizing incompatibility of hypotheses on arthropod phylogeny. According to Lake's results (1990) crustaceans are not closely related to the Tracheata but belong to a group of coelomate protostomians that excludes myriapods and insects. Friedrich and Tautz (1993) prefer a close relationship between insects and crustaceans, excluding myriapods, while according to Wheeler et al. (1993) the Mandibulata (Myriapoda + Insecta + Crustacea) are monophyletic

was written by the second author and will be released soon. Essentially, supporting positions are counted as in Hadamard conjugation (Hendy and Penny 1993), but also considering a low level of noise (autapomorphies in ingroups and convergencies in outgroups). In contrast to other methods that use models of sequence evolution to estimate the number of multiple hits that produce noise and reduce signal, PHYSID is a tool that visualizes patterns without making any assumptions about the mode of sequence evolution. The present version of PHYSID (available on request from the authors) does not identify supporting positions equivalent to synapomorphies when a larger number of autapomorphic substitutions of single species is present; future versions will be improved and will have a higher sensitivity.

The sequences used for spectral analysis of supporting positions were obtained from Genbank. Their species names and accession numbers are: *Mus musculus* X00686; *Oryctolagus cuniculus* X00640, X06778; *Homo sapiens* X03205; *Xenopus laevis* X04025; *Latimeria chalumnae* L11288; *Squalus acanthias* M91179; *Echinorhinus cookei* M91181; *Fundulus heteroliticus* M91180; *Salmo trutta* X98839; *Petromyzon marinus* M97575; *Lampetra aepyptera* M97573; *Branchiostoma floridae* M97571; *Styela plicata* M97577; *Saccoglossus kowalevskii* L28054; *Sphaerechinus granularis* Z37132; *Lipotrapeza vestiens* Z80952; *Ophiomyxa brevirima* Z80953; *Porania pulvillus* Z80955; *Endoxocrinus parrae* Z80951; *Lingula lingua* X81631; *Lygus hesperus* U06476; *Hypogastrura dolsana* Z26765; *Aeschna cyanea* X89494; *Ephemera* sp. X89496; *Lepisma saccarina* X89484; *Scolopendra cingulata* U29493; *Branchinecta packardi* L26512; *Bosmina longirostris* Z22731; *Diastylis* sp. Z22519; *Balanus eburneus* L26510; *Stenocypris major* Z22850; *Ornithodorus moubata* L76355; *Ixodes affinis* L76350; *Odiellus troguloides* X81441; *Androctonus australis* X77908; *Eusimonia wunderlichi* U29492; *Macrobiotus hufelandi* X81442; *Hypsibius* sp. Z9337; *Milnesium tardigradum* U49909; *Euperipatoides leuckarti* U49910; *Porocephalus crotali* M29931; *Enchytraeus* sp. Z83750, U95948; *Eisenia fetida* X79872; *Hirudo medicinalis* Z83752; *Aphrodite aculeata* Z83754; *Lanice conchilega* X79873; *Siboglinum fiordicum* X79876; *Ridgea piscesa* X79877; *Spisula solida* L11266; *Mactromeris polynyma* L11230; *Argopecten irradians* L11265; *Placopecten magellanicus* X53899; *Littorina littorea* X91970; *Helix aspersa* X91976; *Nerita albicilla* X91971; *Scutopus ventrolineatus* X91977; *Liolophura japonica* X70210; *Antalis vulgaris* X91980; *Phascolosoma granulatum* X79874; *Ochetostoma erythrogrammon* X79875; *Priapulus caudatus* X80234; *Prostoma eilhardi* U29494; *Plumatella repens* X81631; *Strongyloides stercoralis* M84229; *Plectus* sp. U61761; *Zeldia punctata* U61760; *Tetracerasta blepta* L06670; *Echinostoma caproni* L065670; *Schistosoma haematobium* Z11976; *Echinococcus granulosus* U27015; *Geocentrophora sphyrocephala* D85089; *Bipalium* sp. D85086; *Moliniformis moliniformis* Z19562; *Pycnophyes kielensis* U67997; *Brachonus plicatilis* U29235; *Gordius aquaticus* X80233, X87985; *Rhopalura ophiocomae* X97158, U58369; *Anthopleura kurogane* Z21671; *Anemonia sulcata* X53489; *Tripedalia cystohpora* L10829; *Mnemiopsis leidyi* L10826; *Scypha ciliata* L10825; *Microciona prolifera* L10827; *Trichoplax adhaerens* L10828; *Sphaeroeca volvox* Z34900; *Diaphonoeca grandis* L10824; *Acanthocoepis unguiculata* L10823; *Fragilaria striatula* X77704; *Thalassionema nitzschioides* X77702; *Rhaphoneis belgicae* X77703; *Ichthyophthirius multifiliis* U17354; *Tetrahymena corlissi* U17356; *Diplodinium dendatum* U57764; *Gymnodinium* sp. AF022196; *Symbiodinium microadriacticum* M88521; *Noctiluca scintillans* AF02200; *Aspergillus niger* X78538; *Penicillium notatum* M55628; *Saccharomyces cerevisiae* J01353.

Sequences were aligned with ClustalW (Thompson et al. 1994), the alignment was corrected by hand to maximize the number of invariable positions. Sequence areas that could not be aligned unambiguously were excluded from the analysis. The alignment was 2856 positions long, of these 917 positions were excluded (234–272, 307–403, 858–1136, 2031–2329, 2684–2764). The remaining alignment still had 1405 variable positions. The program PHYSID (see also Wägele and Rödding 1998) was used to identify split-supporting positions

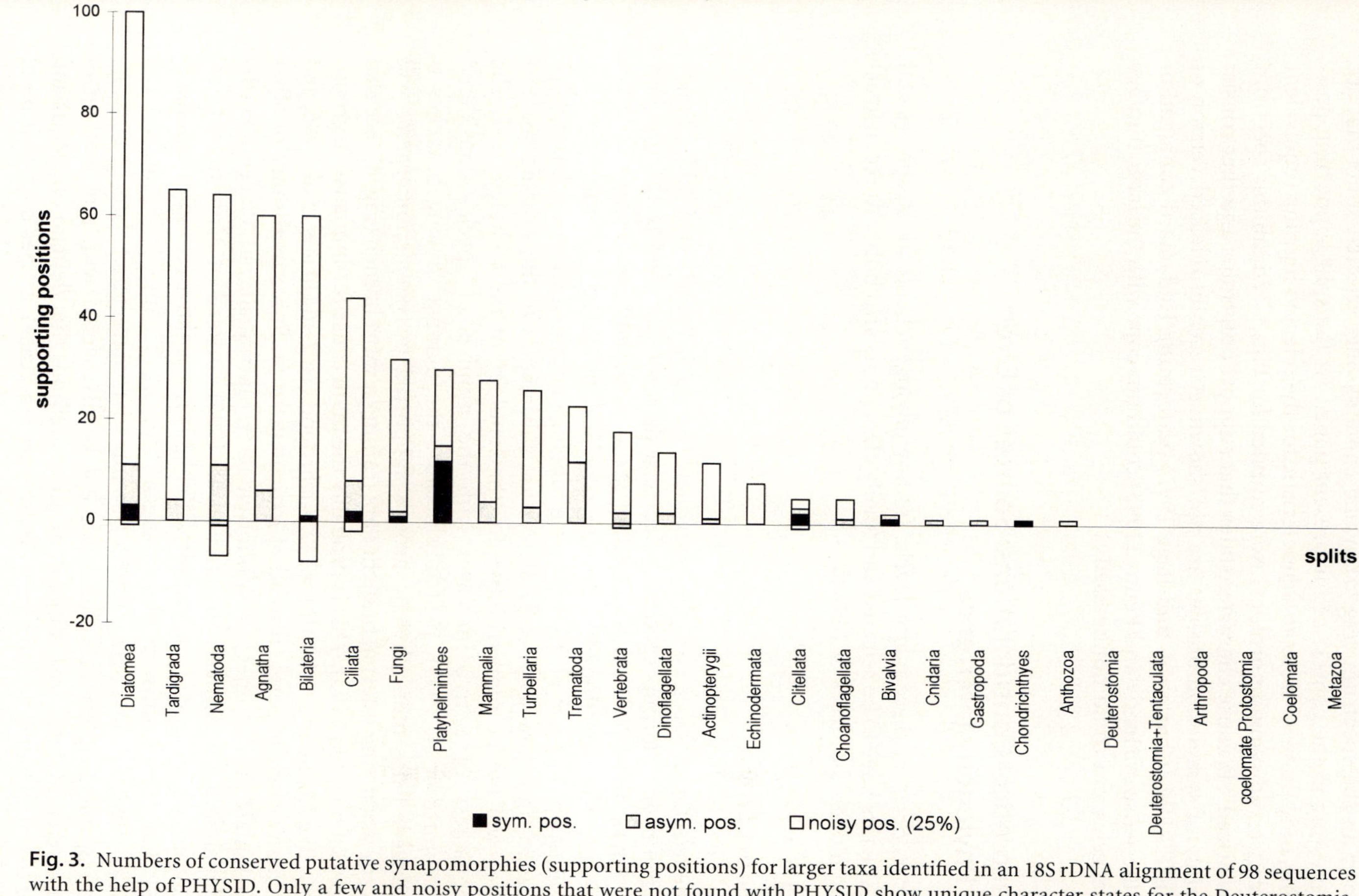

Fig. 3. Numbers of conserved putative synapomorphies (supporting positions) for larger taxa identified in an 18S rDNA alignment of 98 sequences with the help of PHYSID. Only a few and noisy positions that were not found with PHYSID show unique character states for the Deuterostomia, Coelomata, Metazoa and other taxa. Presence of a high signal (large number of supporting positions) means that the taxa are probably monophyletic. *Sym pos.*: symmetric supporting positions (positions with only two character states); *asym. pos.*: positions with more than one character state in outgroup; *noisy pos. 25%:* positions that contain in up to 25% of sequences technical convergencies or autapomorphies (=noise)

shown in the spectrum (Fig. 3), which contains only selected splits that are interesting in the context of the present paper. Of the split-supporting positions, only those with states characteristic for the selected ingroup (equivalent to conserved synapomorphies) were counted for Fig. 3. We call this spectrum "polarized", because character states that support outgroups and that consist mainly of symplesiomorphies are not shown. A recently released version of SPECTRUM (Charleston and Page 1997) performing Hadamard conjugation in the sense of Hendy and Penny (1993) could not cope with the large number of species and could not be used.

3
Theoretical Considerations: Sources of Errors in Phylogeny Inferrence

Errors occur at three major levels: for any phylogenetic analysis, species and characters have to be selected and finally algorithms are chosen to reconstruct homologies and trees or networks.

3.1
Species Errors

The *selection of species* may be inadequate (see e.g. Lecointre et al. 1993). Single ingroup or outgroup species added to a data set can modify the topology of a tree, even if the corresponding taxon is already represented by other species. The selection of outgroup taxa also influences the topology (e.g. Winnepenninckx et al. 1994). A tree topology obtained from a subsample of species (our data sets nearly always are subsamples of the existing species) often will not be representative for the topology of the complete tree. The robustness of a tree increases with the number of species considered (Lecointre et al. 1993), probably because the number of long branches is reduced. False monophyletic groups supported by *symplesiomorphies* will occur when long internal branches are framing a short branch (Fig. 4) (see also Zharkikh and Li 1993). This is probably the main cause for implausible groupings like (Annelida + Mollusca) vs. Arthropoda (e.g. in Giribet et al. 1996, and in many other papers) [instead of Articulata vs. Mollusca in absence of other coelomate protostomes] or (Monotremata + Marsupialia) vs. Placentalia (the *Marsupionta* hypothesis in Janke et al. 1996) [instead of Monotremata vs. Theria], which are not compatible with highly informative (complex) anatomical characters known to morphologists (see e.g. Thenius 1979; Luckett and Zeller 1989; Fox and Meng 1997 for mammalian characters.). This error (clustering due to plesiomorphies) can not be discovered with any method of character analysis, tree statistics or other methods that depend on estimation of tree topologies because they do not vary (increase) species composition of the data set.

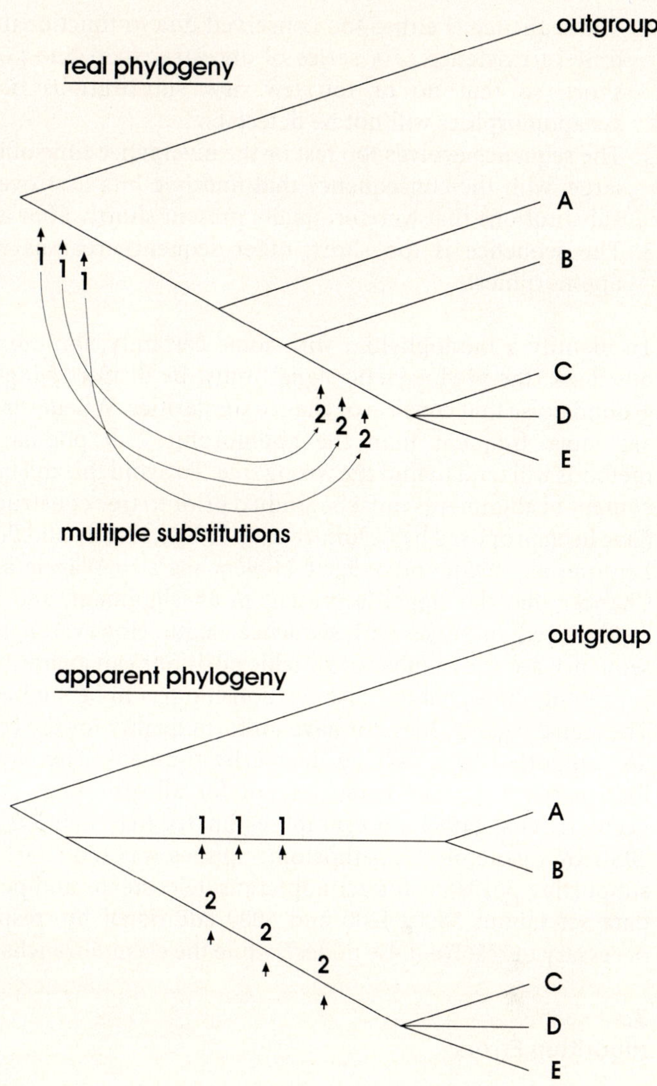

Fig. 4. Diagram illustrating the *symplesiomorphy trap*, a special long-branch problem. Any computer program will find the wrong topology (*below*) with a monophylum A + B when in absence of stem-line substitutions for B-E (real phylogeny, *above*) symplesiomorphies are only conserved in *A* and *B* and masked by multiple hits in *C-E*

3.2
Character Errors

The *selected sequences* may contain *no phylogenetic information*. There are three causes for the lack of information:

1. The sequence is either too conserved due to functional constraints or the time of existence of a series of ancestral stem-line populations was too short, so that no or too few new substitutions occurred. Therefore synapomorphies will not be detectable.
2. The sequence evolves too fast or the divergence time of ingroup-taxa is too large, with the consequence that multiple hits destroyed the apomorphic substitutions that were originally present shortly after a radiation event.
3. The sequence is too short, other sequence regions may contain more apomorphies.

To identify a monophylum with some certainty, the number of supporting positions (the phylogenetic signal) must be distinctly higher than the background noise that consists of chance similarities. Whenever chance similarities are more frequent than the apomorphies, all popular tree-constructing methods will tend to find the wrong tree. To avoid this mistake the information content of alignments must be studied prior to tree construction: new methods have been proposed by Lecointre et al. (1994), Hendy and Penny (1993; see also Lento et al. 1995), and Wägele (1996b, see also Wägele and Rödding 1998). Chances that the signal is present in an alignment, and that noise is more randomized, increases with sequence length. However, neither the length of a sequence nor the number of variable positions can guarantee that information is present; the signal to noise relationship has to be studied for each data set. The signal usually does not have uniform quality for the complete tree; nodes are supported to a varying degree by the data. Therefore, a gene may be informative for some taxa, but not for all groupings contained in a tree. Lecointre et al. (1994) for example estimated that their 532 bp alignment of the 28S-rRNA gene of 38 gnathostome species was too short to contain enough supporting positions for actinopterans, teleosteans and percomorphs in their data set (about 3800, 4200 and 8000 additional bp, respectively, would be necessary to resolve these nodes), while the elasmobranchs were well resolved.

3.3
Algorithm Errors

The *algorithms* used for sequence alignment and tree reconstruction make *assumptions* that may not be realistic. Tree constructing methods require that positional homology has been determined prior to phylogeny reconstruction. Alignment of rDNA sequences is far more difficult than alignment of protein coding genes, where the reading frame imposes an identifiable pattern. Errors caused by suboptimal alignments are rarely examined. Further assumptions of algorithms have to be considered: parsimony methods, for example, require that the probability of homology (= information content) of all characters is similar, a fact that is generally ignored. Distance and maximum likelihood methods need appropriate models of sequence evolution and require that evolution of sequences is a stochastic process.

3.4
Other Sources of Error

The rDNA genes vary within species, so substitutions detected may not be diagnostic characters of a species. Intraspecific variability, a potential source of errors, is rarely studied. The 28S gene, for example, occasionally contains introns in some individuals (5% of the genes of *Locusta migratoria*: Schäfer and Kunz 1985).

The sources of errors mentioned above are responsible for the incompatibilities seen in published dendrograms. Since no reliable procedure for the analysis of data sets is known (new methods like spectral analysis (Hendy and Penny 1993) still have to be tested more intensely), it is reasonable to check the plausibility of hypotheses of relationships with data from other sources (informative morphological characters, biogeography, fossil record).

4
The rDNA Molecules

Metazoan rRNA genes are part of mitochondrial and nuclear genomes. *Mitochondrial genomes* contain two rRNA genes (**12S-rDNA**: about 790–1020 bp, **16S-rDNA**: about 1330–1680 bp) and occur in vertebrates (where, as far as is known, gene order is highly conserved) on the H-strand near the control region, usually framed by single tRNA genes (see e.g. Lee and Kocher 1995). Translocations and inversions can modify rDNA gene order in other taxa (e.g. Clary and Wolstenholme 1985; Smith et al. 1989; Smith 1992; Terrett et al. 1994, Boore and Brown 1994; Boore et al. 1995), events that can be used as apomorphic characters of monophyla. The two rRNA genes are not neighbors in all genomes. Since mtDNA is transmitted maternally in most species, and the complete mitochondrial genome is an individual unit of evolution, it is an important marker for population studies.

Substitution rates vary within the rDNA molecules, being higher in peripheral areas of the secondary structure (e.g. Van de Peer et al. 1993) and on average higher in loops than in stem regions (e.g. Ortí et al. 1996). Of the mitochondrial genes, the rDNAs belong to the more conserved ones. The average impression that mitochondrial genes evolve faster than nuclear genes may not be valid for some taxa (e.g. Brown et al. 1979; Vawter and Brown 1986; DeSalle et al. 1987).

The three larger *nuclear rRNA* genes occur in tandem repeats, with a large number of copies per genome (e.g. about 200 in mammals; more in other vertebrates, thousands in insects: Schäfer and Kunz 1985). The rDNA repeats make up 7% of the genome in *Hermania momus* (Degnan et al. 1990). Each repeat unit has the following structure:

5′NTS---ETS-18SrDNA---ITS1---5.8SrDNA---ITS2---28SrDNA---NTS 3′.

The short nuclear **5S-rRNA** gene (about 120 bp) is usually encoded elsewhere in metazoans and transcribed by another polymerase (RNA polymerase III). It is also present in a large number of clustered copies. However, 5S-rDNA – like sequences have also been found in the rDNA-repeats of *Meloidogyne arenaria* (Nematoda) (Vahidi and Honda 1991) and some arthropods (Drouin et al. 1992). Linkages to other tandemly repeated genes seem to evolve by mechanisms of gene transposition with subsequent homogenization of repeat units (Drouin and Moniz de Sá 1995).

The repeat unit for the larger rRNAs varies in length (e.g. 18 kb long in *Locusta migratoria*: Schäfer and Kunz 1985, between 5 and 9 kb in *Meloidogyne arenaria* (Nematoda): Vahidi and Honda 1991; 7.9 kb in *Herdmania momus* (Ascidiacea): Degnan et al. 1990), most of this variation being caused by the intergenic spacer (**IGS**: length about 4 kb in frogs: Hillis and Davis 1986; 4.8 kb in *Daphnia pulex*: Crease 1993; 3 to 6 kb in *Drosophila*: Coen et al. 1982). The intergenic spacer contains sequences for transcription initiation and termination, and the signals are often contained within subrepeats (see e.g. Crease 1995). Variation in the number of subrepeats is responsible for much of the observed length variation. Most of the IGS is not transcribed (NTS), the last part of the IGS is non-repetitive and contains the external transcribed spacer (ETS, in *Daphnia pulex* 1280 bp, in *Artemia franciscana* 790 bp, in *Drosophila melanogaster* 861 bp, in *Bombyx mori* 909 bp: Crease 1993). Since the IGS evolves fast, this part of the rDNA repeat has not been used for phylogenetic studies of larger taxonomic units.

The **18S-rRNA** gene (= ssu-rDNA) is usually 1800–2000 bp long, but elongations have been observed in several taxa (e.g. Kwon et al. 1991, and our own unpublished results on peracarid crustaceans). The elongations probably are not introns. The **5.8S-rDNA** (about 1600 bp) and **28S-rDNA** (= 1su-rDNA, about 4000–5000 bp) sequences are rarely used in phylogeny reconstruction because the databases are still too small. They are, however, potentially as informative per average nucleotide as the 18Sr-RNA gene.

The internal transcribed spacers (**ITS**; for example ITS1 in *Aedes aegypti*: about 420 bp; ITS2: about 200 bp: Wesson et al. 1992) are not very conserved in insects, showing intraspecific variation (e.g. Wesson et al. 1992; Vogler and DeSalle 1994; Tang et al. 1996), while they might be very conserved in other taxa. The spacers do not evolve neutrally, as indicated by their partly conserved secondary structure (Schlötterer et al. 1994).

Each rDNA gene and the spacers in theory could be independently evolving units; empirical evidence, however, has proved that due to concerted evolution the tandem repeats are homogenized (e.g. Brown et al. 1972; Coen et al. 1982), and the entire repeat with all its components evolves in concert within a species (see review by Elder and Turner 1995). Therefore, polymorphisms are found mostly within, rather than shared among species (Hillis and Davis 1988). Even when polymorphisms within species are not studied, the unnoted autapomorphies of single populations or groups of individuals will not affect phylogeny reconstruction (autapomorphies are not relevant when parsimony methods are used, though they can have an effect on distance methods: Wägele 1996a).

Secondary structure models for the larger rRNAs can be found in the rRNA database of De Rijk et al. (1996). Evolution of secondary structure has rarely been studied. A remarkable example is the elongation in the V3 region of ssu-rRNA seen in cirriped crustaceans (Spears et al. 1994): the evolution of the secondary structure supports conclusions derived from the primary sequences. The D3 region of the lsu-rRNA of isopods proved to be very variable. The secondary structure was estimated by Nunn et al. (1996), but the number of species studied is still too small to allow conclusions on the phylogeny.

5
The Utility of rDNA Sequences Depends on Their Information Content

Sequences must conserve stem-line substitutions (apomorphies) to be useful for phylogeny inferrence. As already mentioned, the rate of evolution and the divergence time of monophyla determine the amount of signal and noise present in an alignment. Usually, conserved sequences are chosen to reconstruct ancient radiations to reduce the number of expected multiple hits. However, a sequence may be conserved, possessing only few variable sites, but nevertheless be too noisy in these positions. It may be that those few positions that are variable accumulate mutations and therefore are too noisy for phylogeny reconstruction. We observe such patterns of high noise levels when comparing arthropods or nematodes with other invertebrates (examples in Wägele and Rödding 1998). Until recently, a phylogenetic signal could not be detected a priori; the fact that nodes were unresolved was used as an indication for the absence of distinct information.

The 12S-rRNA gene is relatively conserved and short, so alignments will, in many instances, contain only a few informative substitutions. Relationships within relatively young taxa can be resolved (e.g. Simon et al. 1990), but reliable reconstruction of phylogeny will not be possible for many highly diverged taxa (Wägele and Wetzel 1994; Wägele and Stanjek 1995). The 16S-rRNA gene varies in a similar way, but it may bear more information because it is longer. Mitochondrial rDNAs have been useful for the study of interspecific relationships between closely related species (e.g. DeSalle 1992; Derr et al. 1992; review in Simon et al. 1994), but for the detection of older divergences (phylogeny of vertebrates, mammals) larger portions or nearly complete mitochondrial genomes have been sequenced to obtain more informative sites (e.g. Zardoya and Meyer 1996; Janke et al. 1996).

For the study of ancient radiations, sequence positions are needed that have evolved so slowly that e.g. paleozoic substitutions have been conserved until today. While absolutely conserved positions are without signal, faster evolving positions are too noisy. Estimations of distances and phylogeny reconstructions suggest that the nuclear ssu-rRNA gene is well conserved in several parts, the 28S-rDNA appears to evolve more rapidly, and the ITS regions are thought to be highly variable and only useful for comparisons between populations and closely related species.

The 18S-rRNA gene is potentially suitable for the study of ancient radiations. Monophyly of taxa of planktonic Foraminifera for example, which diverged from benthic groups of Foraminifera in the Mid-Jurassic, has been documented with ssu-sequences (Darling et al. 1996, 1997). The ssu-rDNA genes conserve some information of Cambrian and Precambrian radiations, but often not enough for definitive conclusions. The divergence between euarthropods and protarthropods (Tardigrada, Onychophora), for example, is probably of late Precambrian age, as both groups are represented with Cambrian fossils. The sequences of tardigrades may allow the assignment to the stemline of euarthropods (Giribet et al. 1996, contradicted by Moon and Kim 1996) with not too great confidence (bootstrap value of 80%, number of supporting positions not identified). 18S-rRNA data of echinoderms have been used to study bifurcations of the larger taxa ("classes") that are about 370 to 555 million years old (Wada and Satoh 1994), while in other studies on smaller groups the same gene is informative only for group divergences within the past 200 million years (Smith 1992). Philippe et al. (1994) studied bootstrap proportions of an ssu-rDNA alignment with 69 sequences and found that major lines of bilaterian taxa could not be resolved. The authors suppose that this molecule is not informative for the identification of stem-lines (i.e. internal lines of a dendrogram separated by nodes) shorter than 40 Myr. There exists, however, no general limit in time for the suitability of rDNA genes or other molecules: the individual history of a taxon leaves different levels of signal and noise in the genes, and each case needs a separate analysis. The rDNA genes of dipterans for example accumulate 3 times as many substitutions per unit of time than other Holometabola (Carmean et al. 1992), and among echinoids the 28S-rRNA gene of echinaceans evolves 3 times faster than in cidaroids or irregular echinoids (Smith et al. 1992).

The 28S-rRNA gene is longer than the 18S gene, and might be of similar use for the study of larger taxa. Early radiation of vertebrates left distinct traces in this gene, however not enough to document each major divergence event (Lecointre et al. 1994). De Rijk et al. (1995) have shown that, by combining mitochondrial and nuclear lsu-rDNA data, plausible trees can be inferred, with monophyletic green plants, animals, fungi, Kinetoplastida, Archaea. The order of divergence of many larger groups is however not well resolved, as indicated by low bootstrap values. Michot et al. (1990) described, for a small portion of the lsu-rRNA, features of the secondary structure that are conserved in larger metazoan groups. There is, however, not enough information to support phylogenies.

The 5S-rRNA gene is too small to be informative for divergent taxa. The dendrograms derived by Hendriks et al. (1986) from 5S-alignments show no meaningful groupings of larger taxa. Halanych (1991) examined 65 sequences of diverse metazoans and could not get meaningful results for the Metazoa, nor for subgroups (Deuterostomia, Mollusca).

6
Monophyly of Metazoans

Early studies of 18S-rDNA or RNA alignments were based on partial sequences and therefore on few variable positions. Results are therefore of questionable value. Field et al. (1988) used about 1000 bp to study "the animal kingdom". The resulting tree contains *polyphyletic* Metazoa, with cnidarians separated from bilaterian taxa. This surprising result certainly is an artifact of the low information content of the data. Some features confirmed by later studies can however be seen: the large distance separating the Bilateria from lower (diploblastic) metazoans, and the basal position of plathelminths within the Bilateria. Other details of the tree topology are probably caused by noise, such as the more basal position of arthropods relative to other spiralian invertebrates, or the placement of the brine shrimp between chordates and echinoderms. Lake (1990), using the same alignment as Field et al. (1988) and the "evolutionary parsimony" method, which relies on transversions (Lake 1987), got results more congruent with morphology (monophyletic Metazoa, Deuterostomia, coelomatic Protostomia), however with many strange details: paraphyletic arthropods, brachiopods between annelids and molluscs, and sipunculids within molluscs.

Evidence in favour of *monophyly* of the Metazoa today dominates over any other hypothesis. Ghiselin (1989) pointed out some potentially apomorphic characters such as a U*U base pair present in the 5S-rDNA gene of all metazoans (positions 80 and 96 in the *Homo sapiens* gene) and a deletion of two nucleotides (positions 1475 and 1476) in the 18S rDNA gene. However, these single observations still have a relatively low probability of homology, and more complex patterns are needed. Lists of potential apomorphies have not been published until now (but see Sect. 8), and bootstrap values and branch lengths have been taken as evidence for the support of a hypothesis of monophyly.

Wainright et al. (1993) used complete 16S-rDNA sequences, and found a close relationship between animals, choanoflagellates and fungi, the animals being monophyletic, with the Choanoflagellata as sistergroup. However, resolution within metazoans was low (as indicated by low bootstrap values and the peculiar position of a bivalve placed close to chordates). Partial 18S-rDNA sequences used by Adoutte and Philippe (1993) and many topologies derived from complete nuclear ssu-rDNA sequences support monophyly of the Metazoa (e.g. Hendriks et al. 1988; Kobayashi et al. 1993; Winnepenninckx et al. 1992, 1994; Cavalier-Smith et al. 1996). De Rijk et al. (1995), studying lsu-rDNA data, obtained a well supported clade for sequences from metazoan mitochondria compared with mtDNA sequences of other eukaryonts. An earlier study by Christen et al. (1991) was based on a less informative alignment of partial 28S-rRNA; with this data set monophyly of metazoans was not obtained. In several studies on single metazoan taxa some protists were used as outgroups, rooting of metazoan topologies then being a byproduct of the analysis. For example, the Metazoa are monophyletic in dendrograms of flatworms, with sponges, cnidarians and *Saccharomyces cerevisiae* as outgroup taxa for the Platyhelminthes (Katayama et al. 1996). In summary, the results derived from rDNA sequences are congruent with morphological data and indicate that the Metazoa are monophyletic (see e.g. Ax 1995).

7
Relationships of Larger Groups of Metazoans

It is not intended to present in the following paragraphs a comprehensive discussion of published literature. Some of the more conspicuous publications are mentioned to show for which larger taxa the rDNA genes were used with some success. Since, in most studies, information content of alignments was not examined, the results are questionable whenever strong conflicts with complex morphological apomorphies occur. The plausibility of the results should be for the moment the criterion for identifying successful reconstructions of phylogeny.

Myxozoa. That the protist-like parasitic Myxozoa might be highly reduced Metazoa was suggested by Smothers et al. (1994). 18S-rDNA sequences of five species fitted to bilaterian sequences in a data set also containing sequences of sponges, cnidarians and of *Trichoplax*. The Myxozoa appear as sistergroup to the classical Bilateria in NJ and MP topologies, single apomorphic characters for the taxon Myxozoa + Bilateria still remain to be identified. The same result was obtained by Cavalier-Smith et al. (1996), who also used 18S-rDNA alignments.

Porifera, Cnidaria, Ctenophora, Placozoa. The split between diploblastic invertebrates and the Bilateria is very distinct in nearly all published rDNA-phylogenies. Few authors examined the relationships between the large diploblastic taxa. Dendrograms estimated for three species of sponges, an anthozoan and some unicellular eukaryotes by West and Powers (1993) are incongruent, with the exception that a choanoflagellate was always the sistergroup to the Metazoa. A recent analysis by Odorico and Miller (1997), who used an rDNA region with the 5.8S and part of the 28S genes (433 bp) could not resolve relationships between Porifera, Cnidaria, Ctenophora and Placozoa. The ctenophore sequence is found with low bootstrap support as sistergroup to the *Trichoplax* sequence. The Placozoa clearly are placed outside the Cnidaria, and within the Cnidaria the branching order is (Anthozoa ((Cubozoa, Scyphozoa) Hydrozoa)). The secondary structure model of the sequenced region contains an interesting A*C pairing occurring only in non-cnidarian diploblastic taxa including *Trichoplax*. This result suggests that earlier analyses, where the diploblastic Metazoans appear to be monophyletic ("Radiata" in Cavalier-Smith et al. 1996) must contain artifacts (e.g. groupings based on plesiomorphies or chance similarities).

Bilateria. Ghiselin noted in 1989 that the Bilateria share a large number of character states of the 18Sr-DNA gene, that must be considered to be apomorphies (see also Sect. 8). Later studies confirmed that the split between diploblastic invertebrates and the Bilateria is very well supported (Winnepenninckx et a. 1992; Adoutte and Philippe 1993; Philippe et al. 1994; Katayama et al. 1995; Garey et al. 1996).

Platyhelminthes. 18S-rDNA alignments of flatworms were recently studied by Katayama et al. (1996). Information content of the alignments was not estimated, reliability of single nodes was derived from boostrap values. With high bootstrap values in MP and NJ topologies, platyhelminths are monophyletic, the Acoela branch off first within the platyhelminths, the Neodermata (Trematoda + Cestoda) are monophyletic (see also Baverstock et al. 1991), and the "Turbellaria" are paraphyletic. Though some differences exist between the branching order obtained from morphology (Ehlers 1985), major features of the molecular analysis are congruent with the hypotheses derived from anatomical and ultrastructural characters. The authors state that information content of 18S-rDNA sequences is insufficient for a more detailed study of turbellarian phylogeny.

Mesozoa. Morphological data suggest monophyly of the taxa of Mesozoa and presence of bilaterian characters (Ax 1995). Ssu-rDNA sequences of dicyemid Mesozoa have been compared by Katayama et al. (1995) with sequences of other Metazoans. A well supported result was not obtained, but the authors favour the hypothesis that the dicyemids are degenerate Bilateria; they appear in the published dendrograms near species of Platyhelminthes and Nematoda, but with very low bootstrap values (<50%). More or less similar results were also obtained by Pawlowski et al. (1996) and Hanelt et al. (1996). Our own observations (Sect. 8) have shown that the 18SrDNA sequence of *Rhopalura ophiocomae* shares a large number of synapomorphies with other Bilateria. Ax (1995) mentions some ultrastructural similarities (structure of ciliary roots) with the parasitic Platyhelminthes.

Nemathelminthes (= Aschelminthes). Recent morphological studies suggest that the Nemathelminthes are not monophyletic (Neuhaus 1994; Ahlrichs 1995). This has been confirmed by 18S-rDNA alignments. Winnepenninckx et al. (1995a) distinguished a clade composed of the Rotifera + Acanthocephala as sistergroup of the Gastrotricha + Platyhelminthes, the Nematoda being at the base of the Bilateria (a long-branch artifact?) and the Priapulida near arthropods. However, this topology seems to be questionable in many parts: bootstrap values are low indicating the presence of a high level of noise. The position of the Nematomorpha could not be assessed. According to Garey et al. (1996) the Acanthocephala are modified rotiferans, and this clade is not related to the Nematoda. Further details of the inferred topology are however not very plausible: the Rotifera and Platyhelminthes are placed within the Coelomata, and priapulids appear as a sister-taxon of arthropods. As in many other cases, reliability of nodes is not estimated with a study of the signal to noise ratio in the alignment. However, the observation that the Rotifera and Acanthocephala are different from other "Nemathelminthes" is congruent with morphology: Ahlrichs (1995, 1997) united these taxa together with the Gnathostomulida in the monophylum *Gnathifera*.

Coelomata. Defining this taxon as being all Bilateria except the Platyhelminthes, Gnathostomulida and the groups united in the artificial taxon "Nemathelminthes" one obtains a large group characterized by a ground pattern with coelome, coelomic gonads, metanephridia, and blood vessels. This group is not seen in some topologies derived from 18S-rDNA alignments. Van de Peer et al. (1993) obtained a tree where platyhelminths occur within the sistergroup of the Deuterostomia. In contrast, Philippe et al. (1994), using a larger data set and after removing fast evolving lineages (nematodes and acanthocephalans) obtained monophyletic Coelomata, supported reasonably well by bootstrap proportions. These data are in accordance with results of other authors (Lake 1987; Winnepenninckx et al. 1992, 1995b, 1996; Adoutte and Philippe 1993; Wada and Satoh 1994;) and congruent with morphology. It seems that the **Nemertini** belong to the Coelomata, but the exact position could not be determined with 18S-rRNA sequences (Turbeville and Field 1992).

Chaetognatha. Chaetognaths are morphologically highly derived, so their systematic position is difficult to assess. Often they are placed within the Deuterostomia (e.g. Brusca and Brusca 1990). According to Wada and Satoh (1994) 18S-rDNA sequences prove that chaetognaths are closer to coelomate protostomes than to deuterostomes (between molluscs and arthropods in their dendrogram). Their alignment however, studied by the present authors, contains only a distinct signal in favour of monophyly of chaetognaths, a high level of background noise and is not informative enough to solve the question of the origin of Chaetognatha. This taxon is obviously highly derived not only in its morphology, but also in its ssu-rDNA gene. Halanych (1996) repeated the 18S-rDNA analysis with other methods and species and found a close association between chaetognaths and nematodes outside the coelomata. Though the author examined whether this grouping is a result of long-branch attraction, it remains to be seen whether there exists a distinct signal in the form of a large number of apomorphic character states.

The Annelid-Mollusc clade. A clade is not necessarily a monophylum, it is a branch of a topology and can be an artifact. This is probably true for the annelid-mollusc clade. The annelid-mollusc clade (or, together with other taxa, the "Eutrochozoa") appear in many molecular phylogenies (e.g. Lake 1990; Winnepenninckx et al. 1996; Garey et al. 1996; Kim et al. 1996), but are rarely accepted by morphologists. A plausible explanation for the implied evolutionary and functional derivation of the molluscan organization from the bauplan of an annelid-like ancestor is not known. Lake (1990; see also Ghiselin 1988) proposed a common ancestor with an open circulatory system and segmentation; but, because segmentation hinders hemolymph flow, such a construction cannot survive. Since annelids and arthropods (the taxon Articulata) share a large number of conspicuous anatomical and developmental synapomorphies (e.g. teloblastic growth, coelomic segmentation, structure of the nervous system, arrangement of coelomic cavities, arrangement of excretory organs, direction of blood circulation), the Onychophora being a link between the annelid "worms" and the armoured euarthropods, the annelidmollusc clade is most likely an artifact caused by symplesiomorphies (see Sect. 3). In comparison with other coelomate protostomes, arthropods are a longbranch taxon with fast evolving lines within the group, so that one has to expect mutliple hits in variable positions. Signal to noise ratios have not been studied in alignments in divergent arthropod sequences.

Interestingly, considering a wider range of "worms", Winnepenninckx et al. (1995b) obtained in their analysis a larger clade with all coelomated protostomes except arthropods and with a sequence of *Lineus* sp. (Nemertini) between annelids and molluscs (NJ tree) or within molluscs (MP tree). The variability of the results depending on the tree-constructing method is an indication of the lack of a distinct signal. The MP tree of this analysis (Winnepenninckx et al. 1995b) even shows arthropods as sistergroup to a clade composed of Deuterostomia + "protostome worms", with many nodes supported by low bootstrap values (<80%). An analysis dedicated to molluscs (Winnepenninckx et al. 1996) even failed to recover the monophyly of the Mollusca, Bivalvia, and Gastropoda. Obviously species composition, alignment quality and information content of the sequences is an important factor also in this subgroup of metazoans. The statement of Winnepenninckx et al. (1995b) that the Pogonophora and Vestimentifera are not annelids contradicts morphological data (Bartolomaeus 1994, 1995) and is not plausible as long as data quality is not estimated. Similarly, information content of the alignment used by Mackey et al. (1996) was so low that close relationship of polychaete taxa could not be recovered in the MP topology; an annelid-mollusc clade is not detected by these authors. Other statements derived from the same alignment ("Ectoprocta are not monophyletic") should be judged critically as long as the signal to noise ratio is not uncovered.

Arthropods. Due to the popularity of Manton's *Uniramia* hypothesis (e.g. Manton 1973), which was based essentially on autapomorphies of arthropod taxa and neglection of the search for homologies (Wägele 1993), polyphyly of arthropods was postulated by many morphologists. Contradicting this view and in accordance with older morphological data (e.g. Snodgrass 1950) ssu-rDNA data support monophyly of arthropods (e.g. Adoutte and Philippe 1993; Van de Peer et al. 1993). Alignments of this molecule are however very noisy when it comes to the study of phylogeny whithin the taxon Arthropoda using few species, probably the result of long and rapid divergence. Without estimation of the number of sequence positions with conserved stem-line substitutions it is risky to infer phylogenies. Dendrograms derived from 12S-rRNA alignments are even less informative for the study of ancient arthropod divergence events. The not very plausible hypotheses that onychophorans are secondarily simplified arthropods related to chelicerates (Ballard et al. 1992) proved to be the result of noise (Wägele and Wetzel 1994).

A common feature of many analyses is the more basal position of the Chelicerata compared with the Crustacea and Insecta, usually in absence of onychophorans and myriapods (e.g. Winnepenninckx et al. 1992). In presence of other arthropods the results are not congruent with morphology. For the study of relationships between myriapods, insects, crustaceans and chelicerates probably longer sequences than hitherto used and a careful analysis of signal and noise levels are needed. Turbeville et al. (1991) and Adoutte and Philippe (1993), for example, obtained with 18S-rDNA sequences a sistergroup relationship between insects and crustaceans, and close relationship between a myriapod (*Spirobolus*) and chelicerates. A similar result was published by Friedrich and Tautz (1995) and Giribet et al. (1996). These clades clearly are not congruent with important morphological characters. The correspondence of different analyses is essentially caused by the use of the same type of data. According to an analysis of the alignment of Friedrich and Tautz (1995) by the present authors, the level of noise is high and only few positions, which might as well be symplesiomorphies or chance similarities, support the clade Myriapoda + Chelicerata. Wheeler et al. (1993), combining ubiquitin and ssu-rDNA sequences calculated a dendrogram with monophyletic Arthropoda, Euarthropoda and Mandibulata, a result concordant with cladistic anatomical analyses. Further data are necessary to clarify this problem.

When concentrating on smaller groups, the 18S-rDNA gene proved to be informative in many cases, the long branch problems are not as effective as in the analysis of all arthropods. A distinct signal in favour of larger groups of the Crustacea (Cirripedia) is present in the alignment used by Spears et al. (1994). The sequence of a member of the Pentastomida clearly is similar to the 18S-rDNA gene of the crustacean *Argulus nobilis* (Abele et al. 1989; Winnepenninckx et al. 1992). On the other hand, Pashley et al. (1993) found that ancient divergence events of holometabolous insects could not be resolved with ssu-rDNA data.

Tentaculata (= Lophophorata). The Tentaculata share with the deuterostomate Pterobranchia radial cleavage, trimeric anatomy, arrangement of coelomic cavities in relation to the lophophore, wherefore the Tentaculata and Deuterostomia are united by some authors in the taxon *Radialia* (e.g. Ax

1995). The Tentaculata can be placed at the base of the deuterostome line, though the blastopore still develops as in the Protostomia. In contrast to these data, dendrograms estimated from 18Sr-DNA alignments studied by Halanych et al. (1995) show the Tentaculata among coelomate protostomes, furthermore the taxon appears to be non-monophyletic. Whether this result is supported by a distinct signal (larger number of potentially apomorphic nucleotides) is not known.

Deuterostomia. In accordance with traditional views, 18S-rDNA data sets clearly support the monophyly of a taxon composed of Enteropneusta, Tunicata (Urochordata), Acrania and Vertebrata (e.g. Holland et al. 1991; Philippe et al. 1994; Garey et al. 1996). Pterobranchs still have to be added to these alignments. Information content of the 18S gene is probably not high enough to resolve conclusively the branching order between Enteropneusta, Echinodermata and Chordata (Turbeville et al. 1994), and published topologies are not well supported (e.g. the clade Hemichordata + Echinodermata in Wada and Satoh 1994).

Vertebrates. Monophyly of Chordata and Vertebrata is a common feature of topologies estimated from ssu-rDNA alignments (e.g. Winnepenninckx et al. 1992; Van de Peer et al. 1993). The major events of the morphological evolution of vertebrates from the Agnatha via different levels of organization of aquatic Gnathostomata to amphibians, tetrapods and the radiation of tetrapods are well understood, implausible results obtained from molecules are therefore easily identified. A consensus maximum-parsimony-tree obtained with partial 28S-rDNA sequences of 31 species of mainly aquatic vertebrates (Lê et al. 1993; Lecointre et al. 1993) contains most major groups in the expected order, the few amphibians however do not form a monophylum, and mammals are placed as sistergroup to the remaining Gnathostomata. Many bootstrap values are low in this study: the sequences are probably too short for the correct placement of several taxa.

We will not go into further details of other studies but stress that dendrograms can be misleading even when high bootstrap values are found (see e.g. "unambiguous support" for a bird-mammal relationship in Hedges et al. 1990; comments by Marshall 1992, and the corrected phylogeny with bird-crocodilian relationship in Hedges 1994).

8
Determination of the Phylogenetical Signal Conserved in 18S-rDNA Sequences

To visualize the amount of conserved phylogenetic information contained in 18SrDNA sequences, we used 98 aligned sequences of diverse metazoans, fungi and protists (see Sect. 2) and the program PHYSID (see Wägele 1996b; Wägele and Rödding 1998). We selected for the spectrum (Fig. 3) only splits with some major taxa, the complete spectrum consists of 5355 splits. The amount of conserved putative synapomorphies is highest for diatoms, tardigrades, nematodes, the Agnatha, and interestingly also for the Bilateria when the sequence of the mesozoan *Rhopalura ophiocomae* is included. For other groups the signal is low, for example for the Vertebrata (lower than for the Mammalia), for the Bivalvia, Cnidaria, and Gastropoda. The absence of signal does not mean that the groups are not monophyletic, it simply indicates that the sequence is not suitable for resolving the corresponding nodes. There is no distinct conserved signal for the Deuterostomia, the Arthropoda, the Coelomata, the Metazoa and other groups. Visual inspection of the alignment proved that PHYSID in its present version excludes positions with putative synapomorphies when positions are too noisy, i.e. when several autapomorphies of single species are present in a single position. For metazoans, for example, seven noisy (less conserved) positions exist in this alignment that clearly have in most species a character state occurring only in metazoans. This indicates that allowing a higher noise level, a signal can be identified also for the metazoans, although it is relatively low in comparison with other taxa. Concerning arthropods, the 18SrDNA sequences do not conserve enough information to resolve phylogeny within this group, at least using the currently available selection of species.

The fact that a very clear conserved signal is present for the Bilateria but none for e.g. the Deuterstomia illustrates that neither the age nor the size of a taxon allows a prediction of the expected amount of conserved phylogenetic signal. The signal depends on the number of stem-line substitutions and the subsequent substitutions within the crown-group that might eradicate part of the signal.

9
Discussion

rDNA genes are popular and in many cases useful, however, molecular systematists tended in the past to believe uncritically in the power of their data and their methods, so that a large number of incompatible hypotheses have been published. These contradictions are symptoms of problems which are not very different from those known to morphologists, the major causes for errors (see Sect. 3) being the same (incomplete species sampling, erroneous determination of homologies, assumptions of tree constructing methods that are not concordant with the true history of the molecules). Lecointre (1996) rightly complained that many systematists do not work with a hypothetico-deductive methodology: the assumptions are not tested, the predictions obtained from mathematical calculations are treated as if they were truths and the plausibility of the results is not questioned. It is a mistake to ignore the enormous body of data collected by morphologists. Morphological characters are often highly informative, because their complexity allows the determination of homology with a high probability that patterns have been identified correctly. The many mistakes published by cladists working with morphological data (see e.g. Wägele 1994) are not caused by inherent problems of morphological characters but by the methods used by scientists and by careless postulations of character identity (see e.g. the "metazoan phylogeny" in Schram 1991).

To get patterns of higher complexity in molecular data we recommend the use of spectral analysis based on the Hadamard conjugation (Hendy and Penny 1993) or the spectra of supporting positions that allow identification of putative apomorphies, as used in Section 8. Thus not only the single positions and nucleotides that support a node in a dichotomous dendrogram can be identified, but also the quality of a signal in comparison with background noise is visualized, without having to use tree constructing methods or any assumption on relationships.

The rDNA genes have the advantage that in some (not all) taxa, variable regions contain information on relatively recent speciation events, while other highly conserved portions of the same gene may show patterns that evolved long ago. Whether such patterns are present should be tested prior to phylogeny reconstruction, because tree constructing methods do not check data quality.

References

Abele LG, Kim W, Felgenhauer BE (1989) Molecular evidence for inclusion of the phylum Pentastomida in the Crustacea. Mol Biol Evol 6:685–691

Adoutte A, Philippe H (1993) The major lines of metazoan evolution: summary of traditional evidence and lessons from ribosomal RNA sequence analysis. In: Pichon Y (ed) Comparative molecular neurobiology. Birkhäuser, Basel, pp 1–30

Ahlrichs WH (1995) Ultrastruktur und Phylogenie von *Seison nebaliae* (Grube 1859) und *Seison annulatus* (Claus 1876) Hypothesen zu phylogenetischen Verwandtschaftsverhältnissen innerhalb der Bilateria. Cuvillier, Göttingen

Ahlrichs WH (1997) Epidermal ultrastructure of *Seison nebaliae* and *Seison annulatus*, and a comparison of epidermal structures within the Gnathifera. Zoomorphology 117:41–48

Ax P (1995) Das System der Metazoa I. Fischer, Stuttgart

Ballard JWO, Olsen GJ, Faith DP, Odgers WA, Rowell DM, Atkinson PW (1992) Evidence from 12S ribosomal RNA sequences that onychophorans are modified arthropods. Science 258:1345–1348

Bartolomaeus T (1994) Ultrastruktur und Entwicklung der Uncini und deren Bedeutung für die Systematik der Anneliden. Verh Dtsch Zool Ges 87:215

Bartolomaeus T (1995) Structure and formation of the uncini in *Pectinaria koreni*, *Pectinaria auricoma* (Terebellida) and *Spirorbis spirorbis* (Sabellida): implications for annelid phylogeny and the position of the Pogonophora. Zoomorphology 115:161–177

Baverstock PR, Fielke R, Johnson AM, Bray RA, Beveridge I (1991) Conflicting phylogenetic hypotheses for the parasitic platyhelminths tested by partial sequencing of 18S ribosomal RNA. Int J Parasitol 21:329–339

Boore JL, Brown WM (1994) Mitochondrial genomes and the phylogeny of molluscs. Nautilus Suppl 2:61–78

Boore JL, Collins TM, Stanton D, Daehler LL, Brown WM (1995) Deducing the pattern of arthropod phylogeny from mitochondrial DNA rearrangements. Nature 376:163–165

Brown DD, Wensink PC, Jordan E (1972) A comparison of the ribosomal DNAs of *Xenopus laevis* and *Xenopus mulleri*: evolution of tandem genes. J Mol Biol 63:57–73

Brown WM, George M, Wilson AC (1979) Rapid evolution of animal mitochondrial DNA. Proc Natl Acad Sci USA 76:1967–1971

Brusca RC, Brusca GJ (1990) Invertebrates. Sinauer, Sunderland, MA

Carmean D, Kimsey LS, Berbee ML (1992) 18S rDNA sequences and the holometabolous insects. Mol Phylog Evol 1:270–278

Cavalier-Smith T, Allsopp MTEP, Chao EE, Boury-Esnault N, Vacelet J (1996) Sponge phylogeny, animal monophyly, and the origin of the nervous system: 18S rRNA evidence. Can J Zool 74:2031–2045

Charleston MA, Page M (1997) Spectrum. Computer program for spectral analysis, available at http://taxonomy.zoology.gla.ac.uk/~mac/spectrum/spectrum.html

Christen R, Ratto A, Baroin A, Perasso R, Grell KG, Adoutte A (1991) An analysis of the origin of metazoans, using comparisons of partial sequences of the 28S RNA, reveals an early emergence of triploblasts. EMBO J 10:499–503

Clary DO, Wolstenholme DR (1985) The mitochondrial DNA molecule of *Drosophila yakuba*: nucleotide sequence, gene organization, and genetic code. J Mol Evol 22:252–271

Coen ES, Thoday JM, Dover G (1982) Rate of turnover of structural variants in the rDNA gene family of *Drosophila melanogaster*. Nature 295:564–568

Crease TJ (1993) Sequence of the intergenic spacer between the 28S and 18S rRNA-encoding genes of the crustacean, *Daphnia pulex*. Gene 134:245–249

Crease TJ (1995) Ribosomal DNA evolution at the population level: nucleotide variation in intergenic spacer arrays of *Daphnia pulex*. Genetics 141:1327–1337

Darling KF, Kroon D, Wade CM, Brown AJL (1996) Molecular phylogeny of the planktic Foraminifera. J Foraminiferal Res 26:324–330

Darling KF, Wade CM, Kroon D, Brown AJL (1997) Planktic foraminiferal moleuclar evolution and their polyphyletic origins from benthic taxa. Mar Micropaleontol 30:251–266

De Rijk P, Van de Peer Y, Van den Broeck I, De Wachter R (1995) Evolution according to large ribosomal subunit RNA. J Mol Evol 41:366–375

De Rijk P, Van de Peer Y, De Wachter R (1996) Database on the structure of large ribosomal subunit RNA. Nucleic Acids Res 24:92–97

Degnan BM, Yan J, Hawkins CJ, Lavin MF (1990) rRNA genes from the lower chordate Herdmania momus: structural similarity with higher eukaryotes. Nucleic Acids Res 18:7063–7070

Derr JN, Davis SK, Woolley JB, Wharton RA (1992) Variation and the phylogenetic utility of the large ribosomal subunit of mitochondrial DNA from the insect order Hymenoptera. Mol Phylogen Evol 1:136–147

DeSalle R (1992) The phylogenetic relationships of flies in the family Drosophilidae deduced from mtDNA sequences. Mol Phylog Evol 1:31–40

DeSalle R, Freedman T, Prager EM, Wilson AC (1987) Tempo and mode of sequence evolution in mitochondrial DNA of Hawaiian *Drosophila*. J Mol Evol 26:157–164

Drouin G, Moniz de Sá M (1995) The concerted evolution of 5S ribosomal genes linked to the repeat units of other multigene families. Mol Biol Evol 12:481–493

Drouin G, Sévigny JM, McLaren IA, Hofman JD, Doolittle WF (1992) Variable arrangements of 5S ribosomal genes within the ribosomal DNA repeats of arthropods. Mol Biol Evol 9:826–835

Ehlers U (1985) Das phylogenetische System der Plathelminthes. Fischer, Stuttgart

Elder JF, Turner BJ (1995) Concerted evolution of repetitive DNA sequences in eukaryotes. Q Rev Biol 70:297–320

Field KG, Olsen GJ, Lane DJ (1988) Molecular phylogeny of the animal kingdom. Science 239:748–752

Fox RC, Meng J (1997) An X-radiographic and SEM study of the osseous inner ear of multituberculates and monotremes (Mammalia): implications for mammalian phylogeny and evolution of hearing. Zool J Linn Soc 121:249–291

Friedrich M, Tautz D (1995) Ribosomal DNA phylogeny of the major extant classes and the evolution of myriapods. Nature 376:165–167

Garey JR, Near TJ, Nonnemacher MR, Nadler SA (1996) Molecular evidence for Acanthocephala as a subtaxon of Rotifera. J Mol Evol 43:287–292

Ghiselin MT (1988) The origin of molluscs in the light of molecular evidence. Oxford Surv Evol Biol 5:66–95

Ghiselin MT (1989) Summary of our present knowledge of metazoan phylogeny. In: Fernholm B, Bremer K, Jörnwall H (eds) The hierarchy of life. Excerpta Medica, Amsterdam, pp 261–272

Giribet G, Garranza S, Baguña J, Riutort M, Ribera C (1996) First molecular evidence for the existence of a Tardigrada + Arthropoda clade. Mol Biol Evol 13:76–84

Halanych KM (1991) 5S ribosomal RNA sequences inappropriate for phylogenetic reconstruction. Mol Biol Evol 8:249–253

Halanych KM (1996) Testing hypotheses of chaetognath origins: long branches revealed by 18S ribosomal DNA. Syst Biol 45:223–246

Halanych KM, Bacheller JD, Aguinaldo AMA, Liva SM, Hillis DM, Lake JA (1995) Evidence from 18S ribosomal DNA that the lophophorates are protostome animals. Science 267:1641–1643

Hanelt B, Van Schyndel D, Adema CM, Lewis LA, Loker ES (1996) The phylogenetic position of *Rhopalura ophiocomae* (Orthonectida) based on 18S ribosomal sequence analysis. Mol Biol Evol 13:1187–1191

Hedges SB (1994) Molecular evidence for the origin of birds. Proc Natl Acad Sci USA 91:2621–2624

Hedges SB, Moberg KD, Maxson LR (1990) Tetrapod phylogeny inferred from 18S and 28S ribosomal RNA sequences and a review of the evidence for amniote relationships. Mol Biol Evol 7:607–633

Hendriks L, Huysmans E, Vandenberghe A, De Wachter R (1986) Primary structures of the 5S ribosomal RNAs of 11 arthropods and applicability of 5S RNA to the study of metazoan evolution. J Mol Evol 24:103–109

Hendriks L, De Baere R, Van Broeckhoven C, De Wachter R (1988) Primary and secondary structure of the 18S ribosomal RNA of the insect species *Tenebrio molitor*. FEBs Lett 232:115–120

Hendy MD, Penny D (1993) Spectral analysis of phylogenetic data. J Classif 10:5–24

Hillis DM, Davis SK (1986) Evolution of ribosomal DNA: fifty million years of recorded history in the frog genus *Rana*. Evolution 40:1275–1288

Hillis DM, Davis SK (1988) Ribosomal DNA: intraspecific polymorphism, concerted evolution, and phylogeny reconstruction. Syst Zool 37:63–66

Holland PWH, Hacker AM, Williams NA (1991) A molecular analysis of the phylogenetic affinities of *Saccoglossus cambrensis* Brambell & Cole (Hemichordata). Philos Trans R Soc Lond [Biol] 332:185–189

Janke A, Gemmell NJ, Feldmaier-Fuchs G, von Haeseler A, Pääbo S (1996) The mitochondrial genome of a monotreme – the platypus (*Ornithorhynchus anatinus*). J Mol Evol 42:153–159

Katayama T, Wada H, Furuya H, Satoh N, Yamamoto M (1995) Phylogenetic position of the dicyemid Mesozoa inferred from 18S rDNA sequences. Biol Bull 189:81–90

Katayama T, Nishioka M, Yamamoto M (1996) Phylogenetic relationships among turbellarian orders inferred from 18S rDNA sequences. Zool Sci 13:747–756

Kim CB, Moon SY, Gelder SR, Kim W (1996) Phylogenetic relationship of annelids, molluscs, and arthropods evidenced from molecules and morphology. Mol Evol 43:207–215

Kobayashi M, Takahashi M, Wada H, Satoh N (1993) Molecular phylogeny inferred from sequences of small subunit ribosomal DNA, supports the monophyly of the Metazoa. Zool Sci 10:827–833

Kwon O, Ogino K, Ishikawa H (1991) The longest 18S ribosomal RNA ever known. Eur J Biochem 202:827–833

Lake JA (1987) A rate-independent technique for analysis of nucleic acid sequences: evolutionary parsimony. Mol Biol Evol 4:167–191

Lake JA (1990) Origin of the Metazoa. Proc Natl Acad Sci USA 87:763–766

Larsen N, Olsen GJ, Maidak BL, McCaughey MJ, Overbeek R, Macke TJ, Marsh TL, Woese C (1993) The ribosomal database project. Nucleic Acids Res 21:3021–3023

Lé HLV, Lecointre G, Perasso R (1993) A 28S rRNA-based phylogeny of the gnathostomes: first steps in the analysis of conflict and congruence with morphologically based cladograms. Mol Phylog Evol 2:31–51

Lecointre G (1996) Methodological aspects of molecular phylogeny of fishes. Zool Stud 35:161–177

Lecointre G, Philippe H, Lé HLV, Le Guyader H (1993) Species sampling has a major impact on phylogenetic inference. Mol Phylog Evol 2:205–224

Lecointre G, Philippe H, Van Le HL, Le Guyader H (1994) How many nucleotides are required to resolve a phylogenetic problem? The use of a new statistical method applicable to available sequences. Mol Phylog Evol 3:292–309

Lee WJ, Kocher TD (1995) Complete sequence of a sea lamprey (*Petromyzon marinus*) mitochondrial genome: early establishment of the vertebrate genome organization. Genetics 139:873–887

Lento GM, Hickson RE, Chambers GK, Penny D (1995) Use of spectral analysis to test hypotheses on the origin of pinnipeds. Mol Biol Evol 12:28–52

Luckett WP, Zeller U (1989) Developmental evidence for dental homologies in the monotreme *Ornithorhynchus* and its systematic implications. Z Säugetierk 54:193–204

Mackey LY, Winnepenninckx B, De Wachter R, Backeljau T, Emschermann P, Garey JR (1996) 18S rRNA suggests that Entoprocta are protostomes, unrelated to Ectoprocta. J Mol Evol 42:552–559

Maidak BL, Olsen GJ, Larsen N, Overbeek R, McCoughey MJ, Woese CR (1996) The ribosomal database project (RDP). Nucleic Acids Res 24:82–85

Manton SM (1969) Evolution and affinities of Onychophora, Myriapoda, Hexapoda, and Crustacea. In: Moore RC (ed) Treatise on invertebrate paleontology, part R: Arthropoda, vol 4. Geological Society of America University of Kansas, pp R15–R57

Manton SM (1973) Arthropod phylogeny – a modern synthesis. J Zool Lond 171:111–130

Marshall CR (1992) Substitution bias, weighted parsimony, and amniote phylogeny as inferred from 18S rRNA sequences. Mol Biol Evol 9:370–373

Michot B, Qu LH, Bachellerie JP (1990) Evolution of large-subunit rRNA structure. Eur J Biochem 188:219–229

Moon SY, Kim W (1996) Phylogenetic position of the Tardigrada based on the 18S ribosomal RNA gene sequences. Zool J Linn Soc 116:61–69

Neuhaus B (1994) Ultrastructure of alimentary canal and body cavity, ground pattern, and phylogenetic relationship of the Kinorhyncha. Microfauna Mar 9:61–157

Nunn GB, Theisen BF, Christensen B, Arctander P (1996) Simplicity-correlated size growth of the nuclear 28S ribosomal RNA D3 expansion segment in the crustacean order Isopoda. J Mol Evol 42:211–223

Odorico DM, Miller DJ (1997) Internal and external relationships of the Cnidaria: implications of primary and predicted secondary structure of the 5'-end of the 23S-like rDNA. Proc R Soc Lond [Biol] 264:77–82

Ortí G, Petry P, Porto JIR, Jegú M, Meyer A (1996) Patterns of nucleotide change in mitochondrial ribosomal RNA genes and the phylogeny of piranhas. J Mol Evol 42:169–182

Pashley DP, McPheron BA, Zimmer EA (1993) Systematics of holometabolous insect orders based on 18S ribosomal RNA. Mol Phylog Evol 2:132–142

Pawlowski J, Montoya-Burgos JI, Fahrni JF, Wüest J, Zaninetti L (1996) Origin of the Mesozoa inferred from 18S rRNA gene sequences. Mol Biol Evol 13:1128–1132

Philippe H, Chenuil A, Adoutte A (1994) Can the Cambrian explosion be inferred through molecular phylogeny? Development (Suppl):15–25

Schäfer M, Kunz W (1985) rDNA in *Locusta migratoria* is very variable: two introns and extensive restriction site polymorphisms in the spacer. Nature 13:1251–1266

Schlötterer C, Hauser MT, von Haeseler A, Tautz D (1994) Comparative evolutionary analysis of rDNA ITS regions in *Drosophila*. Mol Biol Evol 11:513–522

Schram FR (1991) Cladistic analysis of metazoan phyla and the placement of fossil problematica. In: Simonetta AM, Morris SC (eds) The early evolution of Metazoa. Cambridge University Press, Cambridge pp 35–46

Simon C, Pääbo S, Kocher T, Wilson AC (1990) Evolution of mitochondrial ribosomal RNA in insects as shown by the polymerase chain reaction In: Clegg M, O'Brien S (eds) Molecular evolution, vol 122. Liss, New York, pp 235–244

Simon C, Frati F, Beckenbach A, Crespi B, Liu H, Flook P (1994) Evolution, weighting, and phylogenetic utility of mitochondrial gene sequences and a compilation of conserved polymerase chain reaction primers. Ann Entomol Soc Am 87:651–701

Smith AB (1992) Echinoderm phylogeny: morphology and molecules approach accord. Trends Ecol Evol 7:224–229

Smith AB, Lafay B, Christen R (1992) Comparative variation of morphological and molecular evidence through geologic time: 28S ribosomal RNA versus morphology in echinoids. Philos Trans R Soc Lond [Biol] 338:365–382

Smith MJ, Banfield DK, Doteval K, Gorski S, Kowbel DJ (1989) Gene arrangement in sea star mitochondrial DNA demonstrates a major inversion event during echinoderm evolution. Gene 76:181–185

Smothers JF, von Dohlen CD, Smith LH, Spall RD (1994) Molecular evidence that the myxozoan protists are metazoans. Science 265:1719–1721

Snodgrass RE (1950) Comparative studies on the jaws of mandibulate arthropods. Smithson Misc Collect 116:1–85

Spears T, Abele LG, Applegate MA (1994) Phylogenetic study of cirripedes and selected relatives (Thecostraca) based on 18S rDNA sequence analysis. J Crust Biol 14:641–656

Swofford DL, Olsen GJ (1990) Phylogeny reconstruction. In: Hillis DM, Moritz C (eds) Molecular systematics. Sinauer, Sunderland, pp 411–501

Tang J, Toè L, Back C, Unnasch TR (1996) Intra-specific heterogeneity of the rDNA internal transcribed spacer in the *Simulium damnosum* (Diptera: Simuliidae) complex. Mol Biol Evol 13:244–252

Terrett J, Miles S, Thomas RH (1994) The mitochondrial genome of *Cepaea nemoralis* (Gastropoda: Stylommatophora): gene order, base composition, and heteroplasmy. Nautilus Suppl 2:79–84

Thenius E (1979) Die Evolution der Säugetiere. UTB, Stuttgart

Thompson JD, Higgins DG, Gibson TJ (1994) CLUSTAL W: improving the sensitivity of progressive multiple sequence alignment through sequence weighting, position-specific gap penalties and weight matrix choice. Nucleic Acids Res 22:4673–4680

Turbeville JM, Field KG (1992) Phylogenetic position of phylum Nemertini, inferred from 18S rRNA sequences: molecular data as a test of morphological character homology. Mol Biol Evol 9:235–249

Turbeville JM, Schulz JR, Raff RA (1994) Deuterostome phylogeny and the sister group of chordates: evidence from molecules and morphology. Mol Biol Evol 11:648–655

Vahidi H, Honda BM (1991) Repeats and subrepeats in the intergenic spacer of rDNA from the nematode *Meloidogyne arenaria*. Mol Gen Genet 227:334–336

Van de Peer Y, Neefs JM, De Rijk P, De Wachter R (1993) Reconstructing evolution from eucaryotic small-ribosomal-subunit RNA sequences: calibration of the molecular clock. J Mol Evol 37:221–232

Van de Peer Y, Nicolai S, De Rijk P, De Wachter R (1996) Database on the structure of small ribosomal subunit RNA. Nucleic Acids Res 24:86–91

Vawter L, Brown WM (1986) Nuclear and mitochondria DNA comparisons reveal extreme rate variation in the molecular clock. Science 234:194–196

Vogler AP, DeSalle R (1994) Evolution and phylogenetic information content of the ITS-1 region in the tiger beetle *Cicindela dorsalis*. Mol Biol Evol 11:393–405

Wada H, Satoh N (1994) Phylogenetic relationships among extant classes of echinoderms, as inferred from sequences of 18S rDNA, coincide with relationships deduced from the fossil record. J Mol Evol 38:41–49

Wägele JW (1993) Rejection of the "Uniramia" hypothesis and implications of the Mandibulata concept. Zool Jb Syst 120:253–288

Wägele JW (1994) Review of methodological problems of "computer cladistics' exemplified with a case study on isopod phylogeny (Crustacea: Isopoda). Z Zool Syst Evol Forsch 32:81–107

Wägele JW (1996a) Identification of apomorphies and the role of groundpatterns in molecular systematics. J Zool Syst Evol Res 34:31–39

Wägele JW (1996b) First principles of phylogenetic systematics, a basis for numerical methods used for morphological and molecular characters. Vie Milieu 46:125–138

Wägele JW, Rödding T (1998) Estimation of conserved phylogenetic information content of alignments and of probability of homology. Mol Phylog Evol (in press)

Wägele JW, Stanjek G (1995) Arthropod phylogeny inferred from partial 12S rRNA revisited: monophyly of the Tracheata depends on sequence alignment. J Zool Syst Evol Res 33:75–80

Wägele JW, Wetzel R (1994) Nucleic acid sequence data are not per se reliable for inference of phylogenies. J Nat Hist 28:749–761

Wainright PO, Hinkle G, Sogin ML, Stickel SK (1993) Monophyletic origins of the Metazoa: an evolutionary link with fungi. Science 260:340–342

Wesson DM, Porter CH, Collins FH (1992) Sequence and secondary structure comparisons of ITS rDNA in mosquitoes (Diptera: Culicidae). Mol Phylog Evol 1:253–269

West L, Powers D (1993) Molecular phylogenetic position of hexactinellid sponges in relation to the Protista and Demospongiae. Mol Mar Biol Biotechnol 2:71–75

Wheeler WC, Cartwright P, Hayashi CY (1993) Arthropod phylogeny: a combined approach. Cladistics 9:1–39

Winnepenninckx B, Backeljau T, Van de Peer Y, De Wachtler R (1992) Structure of the small ribosomal subunit RNA of the pulmonate snail, *Limicolaria kambeul*, and phylogenetic analysis of the Metazoa. FEBS Lett 309:123–126

Winnepenninckx B, Backeljau T, De Wachter R (1994) Small ribosomal subunit RNA and the phylogeny of Mollusca. Nautilus Suppl 2:98–110

Winnepenninckx B, Backeljau T, Mackey L, Brooks JM, De Wacher R, Kumar S, Garey JR (1995a) 18S rRNA data indicate that aschelminthes are polyphyletic in origin and consist of at least three distinct clades. Mol Biol Evol 12:1132–1137

Winnepenninckx B, Backeljau T, De Wachter R (1995b) Phylogeny of protostome worms derived from 18S rRNA sequences. Mol Biol Evol 12:641–649

Winnepenninckx B, Backeljau T, De Wachter R (1996) Investigation of molluscan phylogeny on the basis of 18S rRNA sequences. Mol Biol Evol 13:1306–1317

Zardoya R, Meyer A (1996) The complete nucleotide sequence of the mitochondrial genome of the lungfish (*Protopterus dolloi*) supports its phylogenetic position as a close relative of land vertebrates. Genetics 142:1249–1263

Zharkikh A, Li WH (1993) Inconsistency of the maximum-parsimony method: the case of five taxa with a molecular clock. Syst Biol 42:113–125

Sponges (Porifera) Molecular Model Systems to Study Cellular Differentiation in Metazoa

W.E.G. Müller, C. Wagner, C.C. Coutinho, R. Borojevic, R. Steffen, and C. Koziol[1]

1
Introduction: Constituent Characters of Metazoa

Evolution is a gradual process whereby primarily new genes are formed either by gene duplication (Ohno 1970) or exon shuffling (Gilbert 1978). New proteins can also be produced by overlapping genes, alternative splicing or gene sharing (Li and Graur 1991). These facts imply that (1) proteins found in a given phylum contain elements or modules which are present already in ancestral protein(s) of members of phylogenetically older phyla and (2) that new combinations of such modules create proteins that possess new functions.

Therefore, it can be postulated that animals which are positioned at the base of the Metazoa – like sponges [Porifera] – are especially rich in molecules displaying functions which they still retain in higher Metazoa, as well as in ancestral molecules comprising modules which in higher phyla evolved to functionally different proteins. The search for such molecules has been successful. As outlined in the following, the structures of the characteristic metazoan genes and their deduced proteins were found to display high homology to sequences from members of higher metazoan phyla. Therefore, it is reasonable to adopt the view that Porifera should be placed in the kingdom Animalia together with the (Eu)Metazoa (Müller 1995, 1998a,b).

In addition, as taken from the first sponge genes, especially that coding for the receptor tyrosine kinase, it is now established that modular proteins, composed by exon-shuffling, are common to all metazoan phyla (reviewed in Müller 1998a). This mechanism of exon-shuffling is apparently absent in plants and protists (Patthy 1995). Consequently, the "burst of evolutionary creativity" (Patthy 1995) during the period of Cambrian explosion which resulted in the big bang of metazoan radiation (Lipps and Signor 1992) was driven by the process of modularization. During this process, the already existing domains were transformed into mobile modules allowing the composition of mosaic proteins (Müller 1997b). The mechanism of modularization is more universal and more versatile – it can be applied to all existing domains – than the process of forming new domains. Therefore, it can be assumed that

[1]Institut für Physiologische Chemie, Abteilung Angewandte Molekularbiologie, Johannes Gutenbers-Universität, Duesbergweg 6, 55099 Mainz, Germany

Progress in Molecular and Subcellular Biology, Vol. 21
W.E.G. Müller (Ed.)
© Springer-Verlag Berlin Heidelberg 1998

during the transition from Protozoa to Metazoa, a process which lasted approximately 1000 million years, the formation of domains with distinct folds was at the center of evolution. After accumulation of a critical number of domains, the mechanism of modularization allowed a rapid formation of a series of mosaic proteins by exon-shuffling.

This gradual process of evolution from Protozoa to Metazoa makes it difficult to outline hard criteria common to all Metazoa. The traditional systematics attempted to define constituent characters of Metazoa (Ax 1995; Reitner and Mehl 1996); (1) multicellularity with cell differentiation, (2) oogenesis with tiny polar bodies, (3) spermatogenesis, (4) omnipotent cells, and (5) collagen. It was claimed that these characters define the status of all multicellular animals as a monophylum; no alternative was assumed (Ax 1995). However, based on recent findings, obtained especially by application of the new techniques in the field of molecular biology, the usefulness of these properties as characters is doubtful. It became known that (1) multicellularity with cell differentiation is seen in a primordial stage already in the green alga *Volvox* (Hallmann et al. 1997). Furthermore, concerning arguments (2) and (3), meiotic division during gametogenesis is very poorly documented in sponges even on the ultrastructural level (reviewed in Simpson 1984) and polar bodies were not seen in at least some sponge species (Kay 1990). The presence of omnipotent cells (perhaps archaeocytes) in Porifera, as originally described (Borojevic 1966), has never been experimentally demonstrated, simply because the establishment of cell culture in sponges could not yet be achieved. The argument of the presence of collagen as a character of Metazoa became weak after the discovery of this structural element in a fungus (Celerin et al. 1996).

Monophyly of Metazoa can be established in a separate approach using the existing molecular data, more specifically amino acid sequence data, from metazoan key molecules as reliable characters. In multicellular organisms, structural molecules have to be expressed in a spatial and temporally ordered manner to form the complex bodyplan. The hierarchical regulatory network of the different somatic cell types within metazoan bodies, whose number has been proposed to increase through time as more complex bodies evolve (Valentine 1994), is controlled by three different qualitative mechanisms (Davidson et al. 1995); (1) a gene expression pattern that generates specific cell lineages through cell specifications (Davidson 1989), (2) pattern formation caused by control genes which define axes during embryogenesis (Holland et al. 1988) and (3) morphogens that contribute to the formation of polarity of an embryo (Slack 1987). Until now, no experimental data have been available to indicate that sponges are provided with genes that assign cells to a defined fate in a given region. However, a cDNA encoding a putative morphogen, the endothelial-monocyte-activating polypeptide, has been recently identified in *G. cydonium* (Pahler et al. 1998).

Key proteins controlling the body plan of multicellular animals are the homeodomain-containing proteins. They function as transcription factors that regulate the expression of genes in temporal, spatial and tissue-specific manner in multicellular organisms (see McGinnis and Krumlauf 1992). During embryogenesis, some larvae of sponges, e.g. the amphiblastulae larvae, show a pronounced polarity (Lévi 1956), suggesting a proximal-distal orientation of the bodyplan. Homeotic proteins from sponges which might be involved in such an axis formation have been described (Degnan et al. 1995). Other homeotic genes are expressed in tissue areas which give rise to organ formation, e.g. *BarH1* transcripts are considered to be transcription factors involved in eye development (Higashijima et al. 1992). Sponges form tissues, e.g. choanocyte chambers (Weissenfels cited in Mehlhorn 1989) and special asexual reproduction bodies, e.g. gemmules (see Simpson 1984), hinting at the existence of homeotic genes controlling organ formation (Carroll 1995). This fact prompted three groups to search for homeotic genes both in freshwater sponges, *Ephydatia fluviatilis* and *Ephydatia muelleri* (Richelle-Maurer et al. 1998; Seimiya et al. 1998) and in marine sponges, *Tethya aurantium* and *Haliclona* sp. (Degnan et al. 1995).

Hence, if those molecules which regulate the cell-cell- and cell-matrix-specific recognition within an organism, e.g. specific cell surface receptors, signal transduction pathways or transcription factors, are homologous within the metazoan kingdom, from Porifera to Chordata, and are not present in Protozoa, then those arguments for monophyly of Metazoa are more solid than those presented in the traditional systematics.

2
Porifera and the Origin of Metazoan Evolution

Sponges (Porifera) are frequently considered to be the simplest multicellular animals, and an intermediate phylum between the kingdom Protoctista and the kingdom Metazoa (Animalia) (Margulis and Schwartz 1995). In view of several distinctive features of sponge body organization, cell differentiation and embryogenesis, they are usually clearly separated from other metazoans and termed Parazoa or Archaemetazoa (reviewed in Willmer 1994), in traditional animal classifications such as proposed by Hyman (1940) and Barnes (1987), and in some modern molecular studies on animal evolution (Nikoh et al. 1997). Another phylum, the Placozoa, has been placed together with the sponges into the subkingdom Metazoa, to separate them from the second animal subkingdom, the Eumetazoa (Nielsen 1995). However, it has been proposed recently that the Placozoa are a sister group of the Cnidaria (Philippe 1994).

Recent molecular data, obtained through sequence analysis of cDNAs and proteins have not given support to the separation of Porifera from Eumetazoa, implying that all metazoans are of monophyletic origin (Müller et al. 1994, 1995; Müller 1995). This conclusion is largely based on sequence data of several genes/cDNAs obtained from the marine demosponge *Geodia cydonium*, coding for adhesion molecules (galectin) and adhesion receptors (integrin receptor, receptor(s) featuring scavenger-receptor cysteine-rich domains), or elements involved in signal transduction pathways (tyrosine kinase receptors, G-proteins, Ser/Thr protein kinases; reviewed in Müller 1997a, 1998a). The presence of common molecular mechanisms for both structural and metabolic cell integration in sponges on one side and in higher metazoans on the other, points towards their close evolutionary relationship and a common pathway of the development of multicellularity in the animal kingdom.

Homeotic genes have been isolated from the freshwater sponges, *E. fluviatilis* and *E. muelleri* (Richelle-Maurer et al. 1998; Seimiya et al. 1998) and also from the marine sponges, *T. aurantium* and *Haliclona* sp. (Degnan et al. 1995). Recently, it was possible to demonstrate that the promoter of the homeotic gene *EmH-3* from *E. muelleri* is active in mammalian cells (Coutinho et al., submitted). For the determination of functional promoter activity, transient transfection experiments in mouse NIH 3T3 cells have been performed; the promoter was fused with the luciferase reporter gene. The data revealed that a 401-nt-long promoter fragment, comprising several binding elements for metazoan transcription factors (Fig. 1A), showed the highest activity, while the 175-bp-long promoter segment, comprising solely the TATA- and Cap boxes, showed only 25% of that activity (Fig. 1B,C). This result was taken as evidence that sponge promoters are activated by factors present in mammalian cells.

During the transition from unicellular Protoctista to multicellular Metazoa, the primary pattern of differentiation implies the presence of at least two different cell types, and the simplest multicellular organism can consist of one cell type specialized for feeding and the other for reproduction (Denis and Lacroix 1993; Wolpert 1990). This is in agreement with the original Roux-Weismann's concept of primary separation of somatic and germinal cell lineages (Weismann 1892), in which the immortal germen produces a mortal soma, which will sustain the growth and reproduction of the organism but will necessarily perish. The concept of programmed senescence and death of the somatic cell lineage is consequently inherent in the concept of early separation of the soma and the germ-cell lineages (Kirkwood and Rose 1991). In view of the proposed monophyletic evolution of Metazoans and the position of sponges at the beginning of the evolution of multicellularity, we address the questions: which molecular mechanisms underlie the evolution of the germ cell and somatic cell lineages, and what is the potential control of their immortality or their programmed senescence and death?

3
Reproduction in Porifera

Sponges reproduce both asexually, by bud and gemmule formation, and sexually by production of gametes (reviewed in Simpson 1984). They lack special reproductive organs, although for some calcareous sponges the "nests" in which the full development from oogonia to the formation of mature swimming larvae takes place have been described (Borojevic 1969).

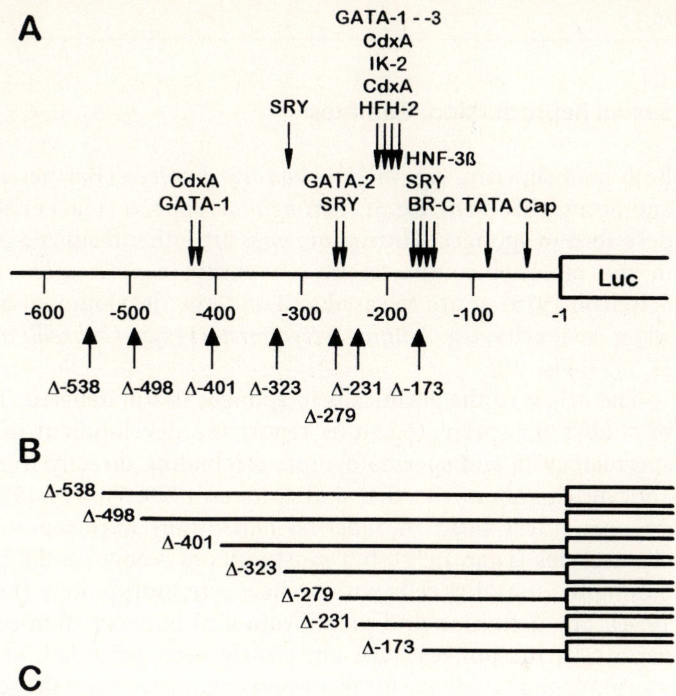

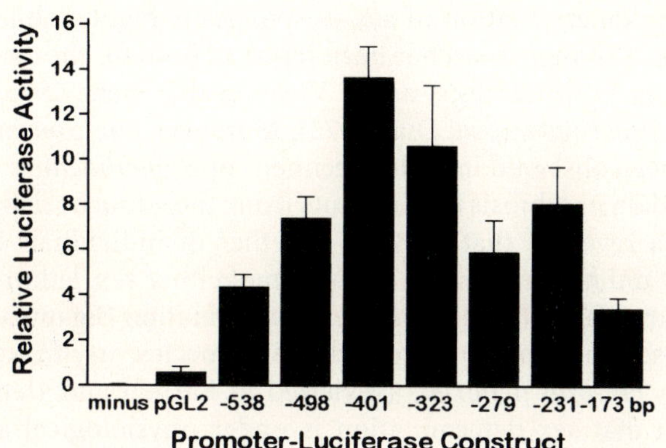

Fig. 1A–C. Structure and differential promoter activity of the *EmH-3* gene from the sponge *E. muelleri*. Promoter-luciferase constructs were used for the analysis. A Schematic representation of the potential binding sites for the transcription factors CdxA, GATA, HFH, SRY, Ikaros [IK-2], BR-C, as well as the TATA box and the Cap site. The numbering starts at the putative transcription initiation site (− for the promoter). *Luc* The luciferase reporter gene fused with the promoter. **B** Stepwise 5′ deletion of the promoter regions upstream of the luciferase reporter gene pGL2. The constructs were named Δ-538, Δ-498, Δ-401, Δ-323, Δ-279, Δ-231, and Δ-173 and refer to the 5′ end of the promoter fragments relative to the transcription start site (+1). **C** The chimeric promoter constructs Δ-538, Δ-498, Δ-401, Δ-323, Δ-279, Δ-231, and Δ-173 were introduced into NIH 3T3 cells together with pCMV-β-galactosidase plasmid and allowed to express for 48h. Controls: pCMV-β-Galactosidase plasmid only (minus) and pGL2-Basic vector (pGL2), containing the luciferase gene, but lacking any promoter and enhancer. Difference in the average magnitude of luciferase expression was normalized for luciferase activity of the pGL2-Control plasmid, which was set to 100%

3.1
Sexual Reproduction: Gametes

Both gonochorism, e.g. in *Halichondria panicea* (Barthel and Detmer 1990), and hermaphrodism, e.g. in *Verongia aerophoba* (Liaci et al. 1971) have been described in sponges. The mature eggs are either fertilized in situ and develop in the mesohyl compartment (viviparity; e.g. *Crambe crambe*; Liaci and Sciscicoli 1975) or are released and undergo development in the surrounding water (oviparity; e.g. *Amoibodictya forsteri* (syn. *Cribrella hamigera*); Zahn et al. 1977; Fig. 2F).

The origin of the germ cells in sponges is still debated. The large majority of studies on spermatogenesis report the development of spermatids from spermatocysts and spermatogonia, originating directly from functional choanocyte chambers (Barthel and Demter 1990; Paulus 1989). The origin of oogonia is less clear: in *Suberites massa* they were reported to derive from choanocytes (Diaz 1979), but most authors recognize the precursors of eggs among the mesohyl cells with archaeocyte morphology (Lévi 1956; Simpson 1984). The definition and identification of putative stem cells for primordial germ cells in sponges have not clearly been provided, and the compelling morphological evidence for the origin of gametes from the somatic fully differentiated cells, such as choanocytes, argues against the clear separation of the germinal and somatic cell lineages.

The determination of sex in sponges is highly labile (Barthel and Detmer 1990). A change in sex has been reported both for the dioecious freshwater (e.g. *Ephydatia fluviatilis*; Van de Vyver and Willenz 1975) and marine sponges (e.g. *Suberites massa*; Diaz 1973). Moreover, the presence of abortive oocytes has been observed in male specimens of *Hymeniacidon sanguinea* (Sarà 1961). Based on parabiosis experiments using the gonochoristic species *Tetilla serica*, which revealed that growing together of individuals belonging to different sexes did not result in an induction of either sex, led to the suggestion that no humoral factor(s) influence sex determination (Egami and Ishii 1956). The sex ratios (female/male) in gonochoristic species vary from 0.6 in *Adocia varia* to 58 in *Petrosia ficiformis* (reviewed in Barthel and Detmer 1990), suggesting again that sex determination is under physiological and not under genetic control (Arndt 1930).

Focusing on the two species of marine Demospongiae which have been most extensively studied at molecular level, *Suberites domuncula* [Metazoa; Porifera, Demospongiae, Tetractinomorpha, Hadromerida, Suberitidae] and *Geodia cydonium* [Metazoa; Porifera, Demospongiae, Tetractinomorpha, Astrophorida, Geodiidae] (Müller 1995, 1997a), the former has been assumed to be hermaphroditic (Arndt 1930), while for the latter species, no data on sexual reproduction have been presented in spite of extensive studies. No conclusive data are available on the life span of these species under natural conditions. Based on the size distribution of the animals collected in the sea, which is reflected by a highly non-symmetrical distribution curve, it can be assumed that *G. cydonium* is a long-living species.

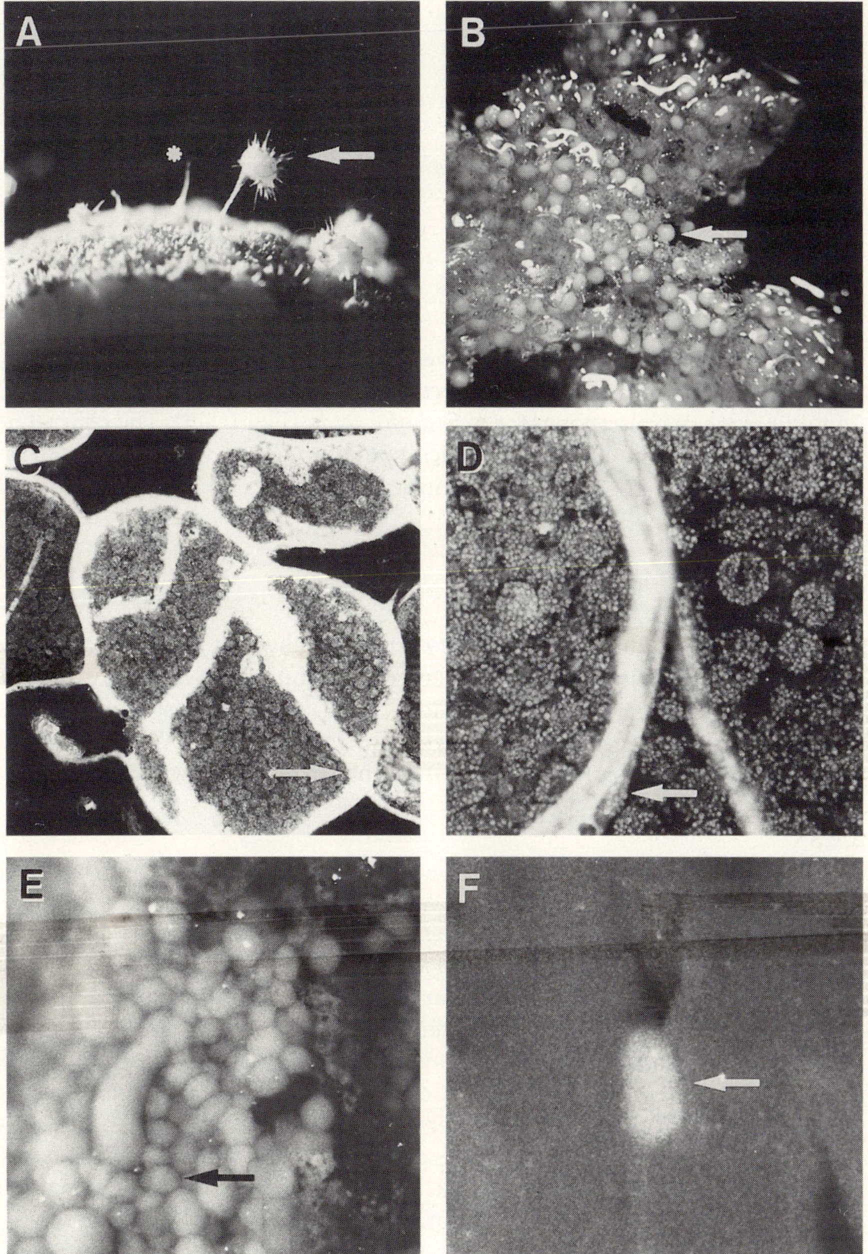

Fig. 2A–F. Forms of reproduction in sponges. **A** Budding in *Tethya aurantium* (*arrow* points to one bud); if the buds drop off they become free and the stalks (*star*) remain with the parent animal (×10). **B** Gemmules (*arrow*) from the freshwater sponge *E. muelleri* (×5). **C–E** Gemmules from the marine sponge *S. domuncula*; cross section through distinct gemmules (**C** at ×25, **D** at ×150) which are surrounded by fused coates (*arrow*) and **E** gemmules present on the outer surface of the shell *Trunculariopsis trunculus* in which the hermit crab *Pagurites oculatus* resides and on which the sponge *S. domuncula* lives; the gemmules (*arrow*) are embedded in their recesses (×6). Gemmule formation in specimens from *S. domuncula* was seen after a 12-day treatment of the animals with heat-killed bacteria, *E. coli*, at a concentration of 10 µg nitrogen ml^{-1}. **F** Egg (*arrow*) released into the surrounding water by the sponge *Amoibodictya forsteri* (×10)

3.2
Asexual Propagation

Sponges have invented several forms of asexual propagation bodies: gemmules, buds and "primordial buds".

3.2.1
Gemmules

Gemmules encapsulate distinct cell types; their size ranges up to 1000 μm. They are very common in freshwater sponges, e.g. *E. muelleri* (Fig. 2B), but are also found in some marine sponges, e.g. *S. domuncula* (Fig. 2C–E). The characteristic feature of gemmules is their thick spongin coat in which microscleres (skeletal elements) are embedded (Simpson 1984).

3.2.2
Buds

In contrast to gemmules, buds are not surrounded by coat layers. They have a complex histology and skeletal arrangement (Bergquist 1978). Usually they are connected with the sponge body by long basal attachment stolons. Typical buds are seen in the marine sponge *Tethya aurantium* (Fig. 2A). After "maturation" the buds are set free and float into the surrounding water.

3.2.3
"Primordial Buds"

Until now, neither gametes, gemmules nor buds, in the defined sense, have been observed in the marine sponge *G. cydonium*. However, at the end of August (Northern Hemisphere), the animals produce specialized fragmentation bodies, which we operationally term here "primordial buds". They protrude from the surface of the specimen but remaining connected by a stalk formed by spicules only (Fig. 3A). Cross sections reveal a zonation of the "primordial buds" into a cortical layer and the medulla (Fig. 3B,C). The only difference between the zones recognized so far, is the size of the sterrasters, which measure 60–80 μm in the cortex, and 40–55 μm in the medulla; no other spicules are found present in the "primordial buds". In the middle of the medulla is one ≈500 μm germinal center, composed of concentrically arranged small cells of a diameter <5 μm (Fig. 3D).

4
Telomerase

So far, there are two major theories to explain cellular ageing: the theory of terminal differentiation and the theory of genetic instabilities (reviewed in

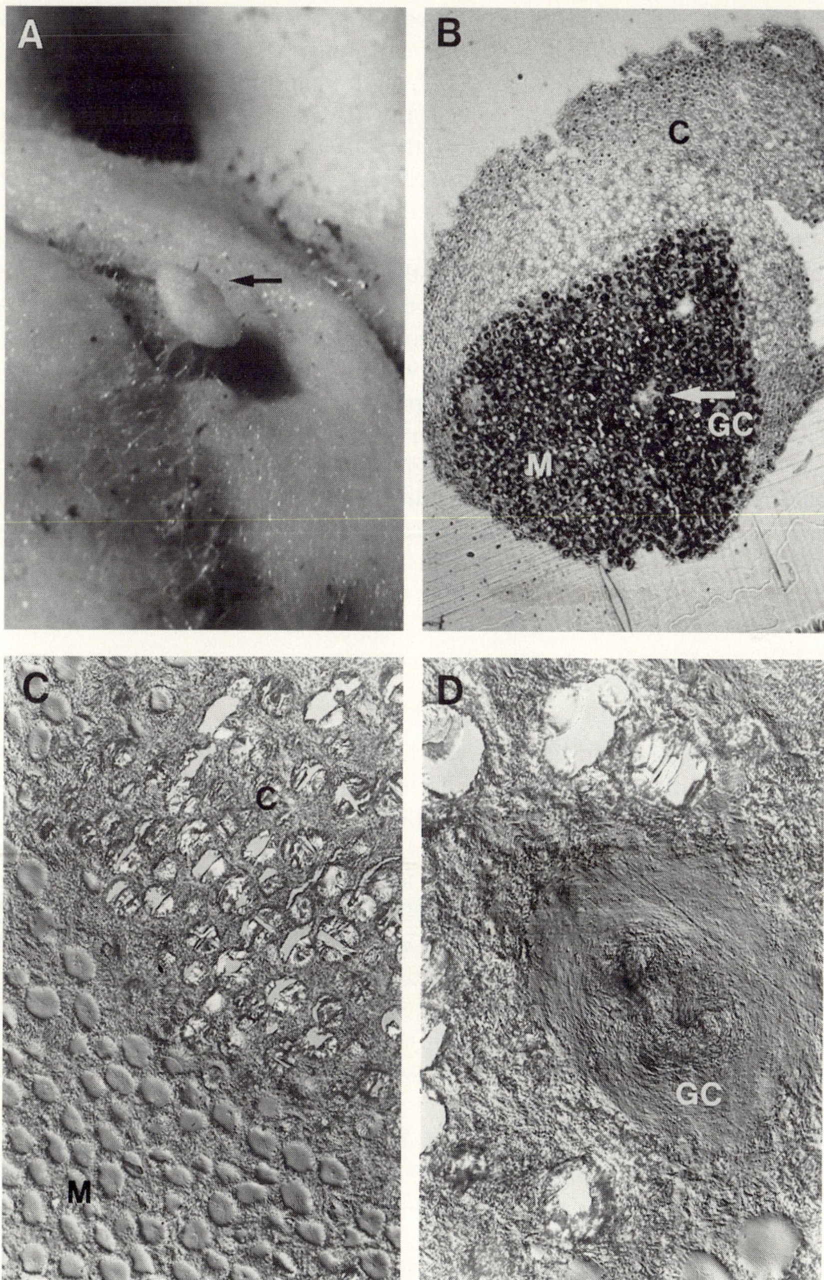

Fig. 3A–D. "Primordial buds" of the sponge *G. cydonium*. A Protrusion of a "primordial bud" (*arrow*) from the surface of the sponge (×1.2). B Cross section through a "primordial bud"; the zonation into the cortical layer (*C*) and the medulla (*M*) as well as the germinal center (*GC*) is marked (×12). C Higher magnification of the two layers, cortex (*C*) and medulla (*M*) (×100). D Germinal center (*GC*) within the medulla (×1000)

Goldstein 1990; Strehler 1986). But recently, ageing has been associated with shortening of the telomeric DNA of chromosomes: it was found experimentally that the length of telomeric DNA in human fibroblasts decreases during ageing as a function of serial passage in vitro and possibly also in vivo (Harley et al. 1990). This hypothesis is attractive because this loss of telomeric DNA is restricted to somatic cells in higher Metazoa (reviewed in Harley 1995), while cells with unlimited replicative potential, such as reproductive cells (Counter et al. 1992) or immortal tumor cells (Hastie et al. 1990) have stable telomers.

The polymerase which mediates the synthesis of additional telomeric repeats was first identified in HeLa cells; it was termed telomerase (Morin 1989). The activity of this enzyme positively correlates with the extent of telomere addition in a given tissue (Harley 1995). In the postnatal somatic tissues in higher metazoan phyla, the reduction in telomerase activity is associated with the programmed senescence and mortality of somatic cell lineages. Hence, the level of activity of telomerase in a given cell population might be used as a parameter to determine its immortality. In order to elucidate whether sponge cells also display these characteristics, the activity of this enzyme was determined in the tissues from *S. domuncula* and *G. cydonium*. Surprisingly high levels of telomerase were detected in these animals (Koziol et al. 1998).

4.1
Telomerase Assay

Telomerase activity in sponges was determined by polymerase chain reaction (PCR) applying the "Telomerase Detection Kit (TRAPeze)" (Kim et al. 1994; Koziol et al. 1998). The products of the telomerase reaction have been visualized as a ladder of oligonucleotides with 6 base increments starting at 50 nucleotides: 50, 56, 62, 68, etc. For a quantification of the relative levels of telomerase activity, the signal intensities of the PCR-amplified telomerase products by a given cell extract were compared with the telomerase activity in a defined number of telomerase positive cells (Chiu et al. 1996). The amount of telomerase activity is given in units of TPG (total product generated) and was calculated as described (Oncor 1996).

4.2
Telomerase Activity in Tissue from *S. domuncula*

The telomerase activity was determined in extracts of tissues from the two sponge species *S. domuncula* and *G. cydonium* by the TRAP assay. The product of the enzyme in the extracts of *S. domuncula* tissue gave a ladder-like pattern, starting from the 50 nt origin with an increment of 6 bases (Fig. 4A,C, lane e). The activity depends on the concentration of the extract. Under the conditions used, the highest activity was found with a concentration of 10 000 cell equivalents in the extract (lane e). Applying lower concentrations from 1×10^1 to 10^3 cell equivalents (lanes d through a; Fig. 4A) the signal intensities of the product

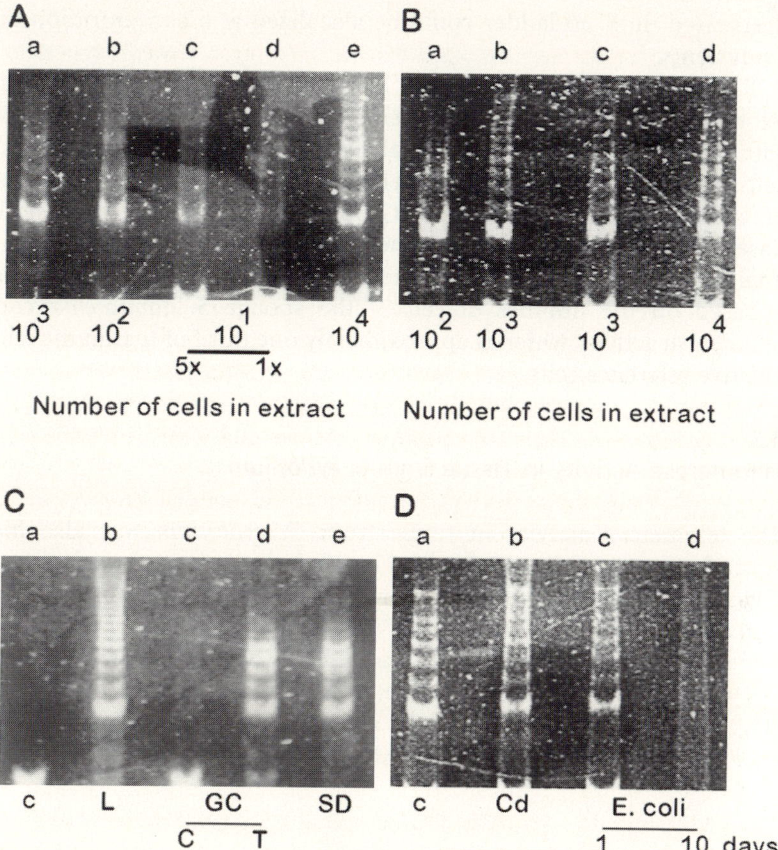

Fig. 4A–D. Telomerase activity in sponge tissue and cells. **A** Detection of telomerase activity in extracts from *S. domuncula* tissue. Defined amounts of tissue, corresponding to the indicated amount of cell equivalents were assayed in the TRAP assay. After PCR amplification, the products were resolved in a non-denaturing polyacrylamide gel and the gels were stained to detect DNA fragments. Different concentrations of extracts, corresponding to 10^4 cell equivalents (*lane e*), 10^3 (*lane a*), 10^2 (*lane b*), 50 (*lane c*) and 10 cell equivalents (*lane d*), were assayed. **B** Analysis of telomerase in tissue from *G. cydonium*. Different amounts of tissue, corresponding to 10^2 (*lane a*) to 10^4 cell equivalents (*lane d*) were analyzed. **C** Comparative analysis of telomerase activity in mouse L5178y leukemia cells and sponge cells/tissue. The concentration applied in the telomerase assay was adjusted to 5×10^3 cells or cell equivalents. Besides mouse cells (*lane b*; *L*), cells and tissue from *G. cydonium* (*lanes c* and *d*; *GC–C* and *GC–T*) as well as tissue from *S. domuncula* (*lane e*; *SD*) were analyzed. A heat-inactivated control (treated at 85 °C for 10 min; *c*) from 5000 telomerase positive reference cells was run in parallel (*lane a*). **D** Telomerase activity in extracts from *S. domuncula* tissue, treated with cadmium or *E. coli* to cause apoptosis. In the experiments the animals remained either untreated (*lane a*; *c*), or were incubated with the cadmium concentration of $5 \mu g\,ml^{-1}$ for 5 days (*lane b*; *Cd*) or with heat-treated *E. coli* for 1 day (*lane c*; *E. coli 1 day*) or 10 days (*lane d*; *E. coli 10 days*). Defined amounts of tissue, corresponding to 5×10^3 cell equivalents, were assayed in the TRAP assay

descreased until no ladder could be visualized at a concentration of 10 cell equivalents.

For the quantification of the enzyme activity, the signal intensity of telomerase-specific products, generated in the TRAP assay, was compared with the signals obtained from serial dilutions of telomerase positive reference cells. As seen in Fig. 5, a correlation between the number of cells and the extent of telomeric additions, given in TPG units, was linear between 10 and 10^4 cells. Based on this calibration curve, the cell equivalents of 5×10^3 cells from *S. domuncula* resulted in an activity of 10.3 TPG units. This result indicates that – based on the number of cells – the sponge *S. domuncula* contains a telomerase activity which is approximately one third of that in the telomerase positive reference cells.

4.3
Telomerase Activity in Tissue from *G. cydonium*

The telomerase activity in tissue from *G. cydonium* was also high. Cell equivalents of 10^2 (Fig. 4B, lane a; Fig. 4C, lane d) showed weak signals of the ladder-like fragments of the telomerase product. Increasing the concentrations to 10^3 (lanes b and c) or to 10^4 cell equivalents (lane d) revealed a strong increase in enzyme activity. A quantitative analysis showed that 5×10^3 cell equivalents from *G. cydonium* contained an activity of 8.3 TPG units (Fig. 5), approximately 20% of the activity seen in telomerase positive reference cells.

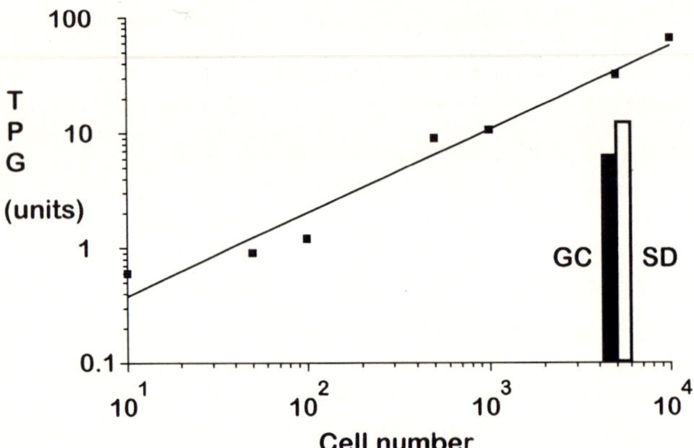

Fig. 5. Calibration curve for telomerase activity using the TRAP assay. The signals obtained from the gels were compared with those (1) from heat-treated samples, (2) from samples with CHAPS lysis buffer only, and (3) from TSR quantitative controls. Telomerase positive reference cells were used to establish the calibration curve. The telomerase activity is given in TPG units. The activities in tissue from *G. cydonium* (*closed bar*; *GC*) or *S. domuncula* (*open bar*; *SD*) corresponding to 5000 cells are indicated

4.4
Telomerase Activity in Cells from *G. cydonium*

In a parallel experiment, dissociated cells from *G. cydonium* were analyzed for telomerase activity. After a chemical dissociation procedure, the spherulous cells were kept for 24h and subsequently analyzed for telomerase activity. The experiments showed that the telomerase activity, if present at all, is lower than 0.5 TPG units (Fig. 4C, lane c).

4.5
Comparison of Telomerase Activity Between Sponges
and Mammalian Tumor Cells

Earlier data showed that immortal cells from mammals contain high levels of telomerase activity (Kim et al. 1994). Therefore, the activities of sponge tissues and cells were compared with those of mouse L5178y leukemia cells. These comparative data showed that at the same concentrations of cells (5000 cells), the activity of mouse L5178y leukemia cells was 227 TPG units (Fig. 4C, lane b), while the activity in the tissues from *G. cydonium* or *S. domuncula* was 8.3 and 10.9 TPG units (lane d and e), respectively.

Isolated cells from *G. cydonium* enriched in the spherulous cell type, did not show any activity (Fig. 4C, lane c). A sample of a heat inactivated control (treated at 85 °C for 10 min) from 5000 telomerase positive reference cells gave only in rare cases a 50-bp signal (lane a).

5
Control of Cell Homeostasis in Sponges: Apoptosis

As mentioned above, it can be assumed that some sponge species, e.g. *G. cydonium*, are long living. The presented studies indicate that the somatic cells in tissue from the two demosponges, *G. cydonium* and *S. domuncula*, species belonging to the widespread families of Geodiidae (Van Soest 1994) and Suberitidae (Arndt 1935), contain high levels of telomerase activity, a feature which is unique in the metazoan kingdom (Koziol et al. 1998). In higher Metazoa the telomerase can only be detected in cells of the germ line (Counter et al. 1992) and in immortal tumor cells (Hastie et al. 1990), while somatic cells are devoid of this enzyme; consequently, they have only a limited potential for proliferation. It can therefore be deduced that the somatic cells of sponges have an unlimited proliferation and differentiation potency (Koziol et al. 1998). However, the bauplan of sponges is, as in all metazoans, a closed one, requiring a control mechanism which allows cell homeostasis, in a sense that the net cell number as the result of cell proliferation and cell death, is well balanced. The process which allows and guarantees such a balance is termed programmed cell death, or apoptosis (reviewed in Thompson 1995).

Two forms of inducers of apoptosis must be distinguished (Buttke and Sandstrom 1994). Firstly, the intrinsic signals, both physiological inhibitors, e.g. estrogens or androgens and physiological activators, like neurotransmitters or calcium; and secondly, extrinsic signals, which can be again inhibitory (viral infection) or activating (heat shock) in their effect on the organism (reviewed in Thompson 1995). In a first rational approach to determine if environmental hazardous compounds cause apoptosis in sponges, the marine sponge *G. cydonium* was used. Previously, it was described for the first time for sponges in particular and for Metazoa in general that xenobiotics, e.g. tributyltin and methyl mercury, cause apoptosis (Batel et al. 1993).

5.1
Induction of Apoptosis in Sponges

Apoptosis was measured by DNA fragmentation assays, applying both the agarose gel electrophoresis technique and the centrifugation method (Müller et al. 1992). The induction of this process was further substantiated by application of a molecular probe, the *MA-3* cDNA (Shibahara et al. 1995), isolated from the same demosponge *S. domuncula*. The application of a molecular marker for the identification of apoptosis in the sponge *S. domuncula* was essential, because it was found that extrinsic (cadmium-induced) apoptosis in *S. domuncula* leads to no death, while intrinsic apoptosis leads to gemmule formation. The latter process in *S. domuncula* was induced by feeding them with the heat-killed bacteria, *Escherichia coli* (Rasmont 1963).

5.1.1
Cadmium-Induced Apoptosis

Cadmium is known as a hazardous metal in the marine environment, and is found as a contaminant in the animals of the human food chain in general and in seafood in particular (Clark 1997). The model compound cadmium – as $CdCl_2$ – was used in the experiments at concentrations which are found in the aqueous environment at polluted sites (Mutwakil et al. 1997). Specimens from *S. domuncula* were exposed to $CdCl_2$ concentrations ranging from 0.01 to $5.0\,\mu g\,ml^{-1}$ for up to 5 days. The extent of DNA fragmentation was determined quantitatively by application of the DNA sedimentation technique. The experiments show that a significant increase in DNA fragmentation was measured at a cadmium concentration of $0.01\,\mu g\,ml^{-1}$. With increasing concentrations, the extent of DNA fragmentation increases and reaches the value of 45.7% at a concentration of $5.0\,\mu g\,ml^{-1}$.

In order to determine the size of the DNA fragments degraded during the process of apoptosis, the DNA was separated by electrophoresis in an agarose gel. As seen in Fig. 6A (lane b), the characteristic ladder-like pattern of DNA with degradation products in multiples of $\approx 180\,bp$ was detected. Such a pat-

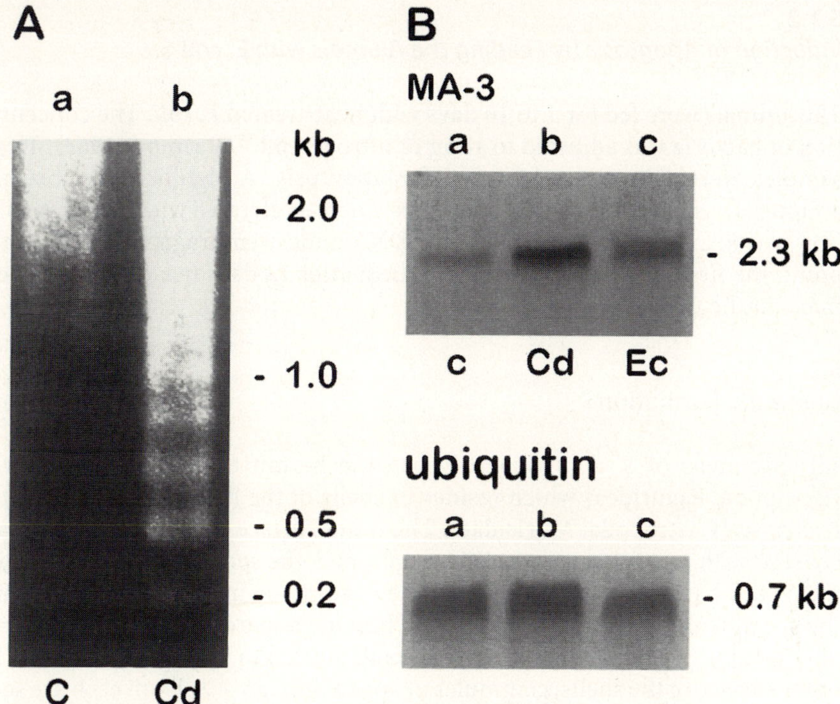

Fig. 6A,B. Apoptosis in *S. domuncula*: DNA fragmentation and expression of *SDMA3* gene. A Cadmium-induced apoptotic fragmentation of DNA in tissue from *S. domuncula*. The specimen remained either untreated (*lane a; C*) or was treated with 0.3 μg ml^{-1} cadmium for 5 days (*lane b; Cd*). DNA was extracted from the cultures and analyzed by agarose (1.2%) gel electrophoresis. The sizes of the DNA fragments are indicated. B Expression of *S. domuncula SDMA3* gene, as analyzed by Northern blotting. Tissue from (1) a control animal that remained untreated (*lane a; c*), (2) a cadmium-treated specimen (exposed for 5 days to 0.1 μg CdCl$_2$ ml^{-1}, *lanes b; Cd*) or (3) an animal fed for 10 days with heat-treated *E. coli* (*lanes c; Ec*) has been used for RNA extraction. 7.5 μg mRNA each were electrophoresed, blotted, and hybridized to either the *SDMA3* probe or to the ubiquitin probe from *S. domuncula*. The signals obtained with the ubiquitin probe show that almost equal amounts of RNA were applied to the gel

tern is characteristic for apoptotic DNA fragmentation (Wyllie 1980). In the controls (absence of CdCl$_2$) no fragmentation was observed (Fig. 6A, lane a).

The concentration of 1 μg ml^{-1} was selected to determine the time of onset of apoptosis in tissue from *S. domuncula*. This time kinetic experiment revealed that after 24h, a significant increase in DNA fragmentation occurred; after 120h, a value of 63.1% for DNA fragmentation was reached if tissue samples from the cortex region were analyzed. The extent of apoptosis in the inner, medulla region is less pronounced.

Under these conditions, none of the animals died and none of them was seen to start gemmule formation.

5.1.2
Induction of Apoptosis by Feeding the Animals with E. coli

The animals were fed for 1 to 10 days with heat-treated *E. coli*. The concentration of bacteria was adjusted to 10 μg of nitrogen ml^{-1} (Rasmont 1963). Tissue samples were taken and analyzed for apoptosis. A significant increase in apoptosis was seen after 1 day, a process which accelerated with longer incubation periods. After 10 days, 36% of the DNA underwent fragmentation. At this point, the first animals in the culture died; after 12 days five of the ten specimens in the experiment had been lost.

5.2
Gemmule Formation

All specimens of *S. domuncula* live on the hermit crab *Pagurites oculatus* [Decapoda: Paguridea] which resides in shells of the mollusc *Trunculariopsis trunculus* [Gastropoda: Muricidae]. The outer surface of these shells is totally covered with 1×1 mm recesses. As mentioned, the sponges died after feeding them with bacteria. After removal of the dead tissue from the shells on which the specimens of *S. domuncula* lived, it became apparent that during the ≈ 12 days feeding period with *E. coli*, the animals started to form gemmules. On the outer surface of the shells, gemmules of approximately 1×1 mm each are seen, which reside in the recesses described (Fig. 2E). Cross sections through the gemmules revealed that they are surrounded by a 40-μm-thick outer layer, which is optically dense. In the interior of the capsule-like gemmules, cells with a uniform shape are present; they have an average diameter of 30 μm (Fig. 2C,D).

5.3
Induction of Expression of *SDMA3* Gene

The cDNA, *SDMA3*, encoding the putative MA3_SD protein was isolated from a library; the cDNA obtained was 2247 nt long. Northern blot analysis was performed with the sponge *SDMA3*-clone as a probe. One band of approximately 2.3 kb was obtained (Fig. 6B), confirming that the full-length cDNA was isolated. Homology searches with *S. domuncula* MA3_SD revealed highest similarity to the mouse MA-3 protein described by Shibahara et al. (1995; Fig. 7). This protein was found to be induced during the apoptotic process (Fig. 6B). The sponge protein sequence shares 40% of identical amino acids and 57% of similar amino acids with the corresponding mouse molecule.

Like the mouse gene (Shibahara et al. 1995), the sponge MA-3 gene undergoes increased expression in response to the apoptotic stimuli. If the sponges are treated with cadmium at a concentration of 0.1 μg ml^{-1} for 5 days, a 15-fold increase – with respect to the control level – of the 2.3-kb transcript can be

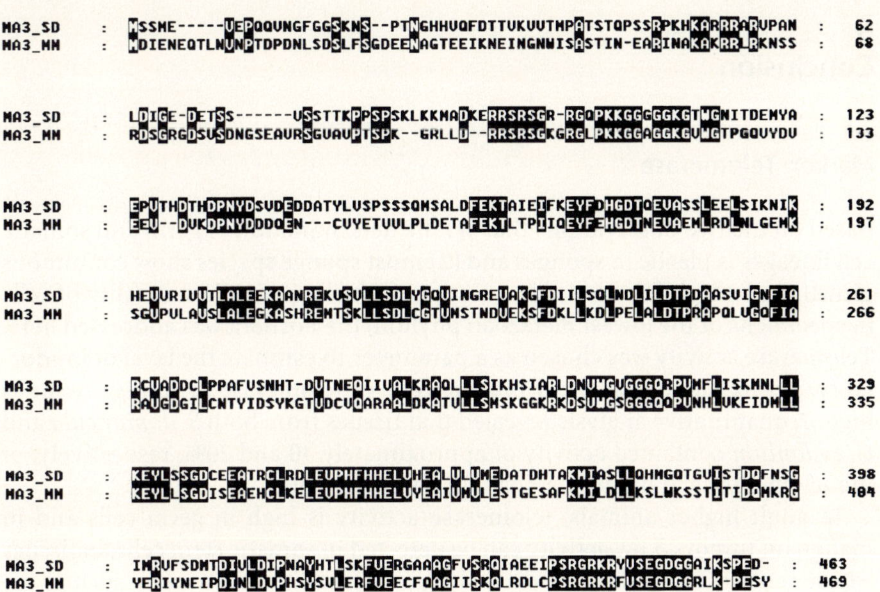

Fig. 7. Alignment of the deduced aa sequence from the sponge MA3_SD protein, and the corresponding mouse MA3-protein (MA3_MM). Residues conserved (identical) among the two sequences are shown in *inverted type*

measured (Fig. 6B; lane b versus lane a). Also, after feeding the sponge with heat-treated *E. coli*, the specimen reacts with a 5.8-fold higher expression of MA-3 (Fig. 6B, lane c).

5.4
Telomerase Activity in Tissue from *S. domuncula* in Response to the Apoptotic Stimuli

As described earlier (Sect. 4.2) untreated control tissue from *S. domuncula* showed a ladder of oligonucleotides with 6 base increments starting at 50 nucleotides: 50, 56, 62, 68, etc. (Fig. 4D, lane a). If the sponges were treated with high cadmium concentrations for 5 days, the activity dropped to 7.5 TPG units (Fig. 4D, lane b) a value which did not change during a longer incubation period. In contrast, the enzyme activity in tissue from animals fed with heat-treated *E. coli* decreases drastically; after 1 day the activity dropped to 7.1 TPG units (lane c), while after 10 days, no activity at all can be detected (lane d).

6
Conclusion

6.1
Marker: Telomerase

Based on the two observations that (1) the determination of germ and somatic cell lineages is plastic in sponges and (2) most sponge species show continuous growth and a long life span, the question of the degree of immortality of cells in specimens of the lowest metazoan phylum, the Porifera was addressed here. Telomerase activity was chosen as a parameter to estimate the level of immortality. It was found that the overall activity of telomerase in sponge tissue is high. A quantitative analysis revealed that tissues from both *S. domuncula* and *G. cydonium* contained activity of approximately 30 and 20%, respectively, of the telomerase activity in the positive reference cells.

In adult higher animals, telomerase activity is high in germ cells and in malignant tumors. Low activity can be detected in somatic stem cells, in tissues whose renewal depends upon extensive proliferation of these cells, such as the hematopoietic system, the bulb-containing fragment of the hair follicles, some epithelia, and germinative centers of the lymph nodes (Norrback et al. 1996; Yasumoto et al. 1996; Ramirez et al. 1997). In these somatic stem cells, this low telomerase activity is transient and responsive to growth factors (Engelhardt 1997). Conversely, in immortal and maturation-sensitive cell lines, the induction of quiescence by contact inhibition, growth factor removal or induction of terminal differentiation, represses the activity of telomerase (Albanell et al. 1996; Holt et al. 1996). In the studied sponges, the number of germ cells in tissues is very low or nil, and the observed high telomerase activity may indicate either that somatic sponge cells maintain the telomerase activity and have an unlimited replication potency, or the sponge tissues contain a large number of somatic stem cells able of unlimited proliferation, undergoing terminal differentiation subsequently. In Cnidaria, a diploblastic group related to the Porifera, three self-renewing stem cell lineages with unlimited capacity for proliferation are in a steady state, and can produce all the differentiated somatic cells, which are continuously shed from the tentacles or included in the growing buds (Bode 1996). In sponges, archaeocytes are considered to be pluripotent, equivalent to stem cells, and potentially able to produce all the major cell lineages, but morphological data indicate that tissue homeostasis is normally maintained by proliferation of all major cell types (Borojevic 1966, 1970). The plasticity of sex determination in sponges, and the ability of fully differentiated cells such as choanocytes to give rise to gametes, favors the hypothesis of potentially high permanent telomerase activity in sponge somatic cells.

Since sponges are metazoans with integrated multicellular organization, they should be provided with mechanisms to control both the rate of cell proliferation and the rate of terminal differentiation or cell death, in order to allow a maintenance of homeostasis of the relative number of different cell types in tissues as well as the absolute number of cells in a given specimen. From earlier studies it is known that single cells from *G. cydonium* that lack any contact with extracellular adhesion factors stop cell proliferation (Gramzow et al. 1989). Telomerase activity is repressed in cells that exit the cell cycle (Holt et al. 1996), and our observation that isolated sponge cells in suspension have an undetectable level of telomerase activity fits well with cell growth inhibition. This is distinctly different from the fully transformed and maturation-resistant human cancer cells, which have no anchorage dependence and maintain both proliferation and telomerase activity in suspension (Albanell et al. 1996).

6.2
Marker: Apoptosis

It is interesting that during an incubation period of 5 days in cadmium-containing seawater, the animals respond with induction of apoptosis. This process was measured in *S. domuncula* both by biochemical and molecular biological techniques. It was demonstrated that (1) the DNA of cadmium-treated animals showed the characteristic ladder-like fragmentation pattern, with fragments in multiples of ≈ 180 base pairs (Wyllie 1980), and (2) apoptotic degradation of DNA was determined by the application of the centrifugation technique. In addition, as a second independent approach, a molecular probe characteristic for apoptosis was applied. The mouse *MA-3* gene is known to be expressed in several mammalian tissues after induction of apoptosis (Shibahara et al. 1995); 6h after inducing the apoptotic process, this gene is upregulated in mammalian tissue. Also, in the *S. domuncula* system, the *MA-3* gene is induced in response to cadmium, as an inducer of apoptosis.

Apoptosis is not only a pathophysiological but also a physiological process which is required in Metazoa for a controlled development of an organism, especially during embryogenesis (Bowen and Bowen 1990). Therefore, it was essential to determine if this process of programmed cell death is involved in sponges in pathway(s) functioning during reproduction, either by gamets or ramets. As mentioned, the determination of sex in sponges is highly labile. A change in sex has been reported both for the dioecious freshwater and marine sponges (e.g. *Suberites massa*; Diaz 1973). As mentioned above, sex determination is under physiological and not under genetic control in sponges. In the present study, the formation of gemmules was induced by feeding the animals with the heat-killed bacterium *E. coli* as described earlier (Rasmont 1963). Under the conditions used, the treated sponge specimens formed gemmules after an incubation period of 10 to 12 days. Tissue samples from animals undergoing gemmule formation showed pronounced apoptosis. This finding underscores again the importance of apoptosis for the maintenance and the survival of multicellular organisms.

6.3
Shift from Immortal to Senescent Cells: Telomerase Activity as a Marker

The apparently broad positive and negative regulation of telomerase activity in sponge cells indicates that they may be a useful model for studying the molecular mechanisms of senescence controls. It has recently been summarized (Vojta and Barrett 1995) that somatic cell growth arrest in Metazoa might be controlled by a complex mechanism driven by multiple senescence pathways. According to Harley (1991; Harley et al. 1994), somatic cell senescence can be subdivided into two phases (Fig. 8). The "Mortality Phase 1" (M-1) leads cells to a permanent cell cycle senescence arrest, a checkpoint at which cells do not respond to any growth factor stimulation. In a second process, the telomeres of the "precrisis" cells reach a critical length, the cells enter "Mortality Phase 2" (M-2) and are prone to the signal for cell death. A series of questions remain open. One major point is the elucidation of whether sponges have the clock that initiates the switch from immortal cells to "somatic cells" committed to

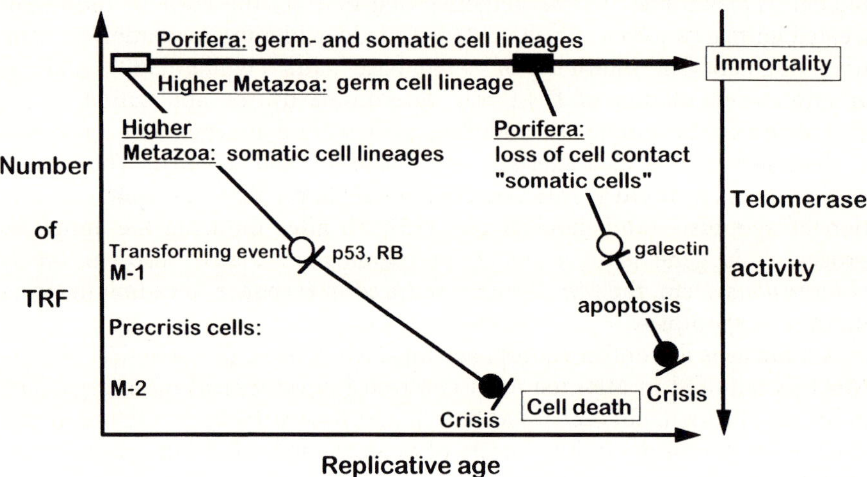

Fig. 8. Hypothetical model for telomere loss and its consequence, the final life span, in metazoan cells. It has experimentally been proven that the cells of the germ line in higher Metazoa and the cells of both the germ and the somatic lineage in Porifera contain high levels of telomerase, thus allowing immortality. In higher Metazoa, an early loss of telomerase activity determines the fate of the somatic cells to senescence (*open box*) via the two phases; (1) the "Mortality Phase 1" (*M-1*) resulting in cell cycle arrest and (2) the "Mortality Phase 2" (*M-2*), which is initiated after transformation. In sponges it is proposed that the switch from immortal somatic cells to mortal "somatic" cells occurs after activation of an endogenous trigger. The M-1 point in sponges is assumed to be reached after only a few rounds of cell replication. The second phase towards cell senescence is supposed to be induced by central death signal(s), e.g. activation of the expression of the *MA-3* gene. The apoptotic process initiated by exogenous signals, e.g. cadmium, is assumed not to result in a shift of the immortal to the mortal stage of the cells. The reduction of telomeres is given in terminal restriction fragments (*TRF*). (Adapted from Harley 1991; Koziol et al. 1998)

differentiation and designated to undergo senescence. Furthermore, the question has to be addressed as to whether and during which phase the reduction of telomeres in the terminal restriction fragments [TRFs] in sponge chromosomes occurs, and whether this reduction concerns a predetermined cell subpopulation, irreversibly committed to differentiation, such as observed in the hematopoietic system, or whether it is stochastic and potentially reversible under specific signals, such as induction of gametogenesis.

The presented data reveal that *S. domuncula* undergoes apoptosis induced by heavy metals, here cadmium, without dying and also without any sign of gemmule formation. The same process is seen if the animals are treated with killed bacteria, very likely due to malnutrition (Rasmont 1963). The latter apoptotic process is paralleled with a reduction of telomerase activity. This important finding resembles in an obvious manner findings in higher animals. As an example, induction of the process of differentiation-induced apoptosis in immortal cells by inducers, e.g. dimethyl-sulfoxide, causes a reduction of telomerase activity (Bestilny et al. 1996). Projected to the sponge model, the cells which appear to be immortal and telomerase-positive undergo apoptosis during the process of "differentiation", during gemmule formation. Those cells not involved in the assembly of gemmules become mortal and die. Hence, according to the data presented, apoptosis in sponges is the result of two different pathways, one which is initiated by exogenous factors, e.g. heavy metals, and does not result in a change of telomerase activity and a second one, caused by endogenous factors, which leads to gemmule formation and a shift of the presumably immortal cells to mortal ones (Fig. 8).

Furthermore, the presented data indicate that somatic sponge cells have a high telomerase activity, which may be controlled by extrinsic factors such as contact with other cells or extracellular matrix. Sponge cell proliferation is under control of growth factors, and the required major pathways of both tyrosine and serine/threonine kinases for intracellular signaling have been described (Ottilie et al. 1992; Schäcke et al. 1994; Kruse et al. 1996). It will be interesting to find out whether these pathways may stimulate and maintain the high telomerase activity in somatic cells. Alternatively, morphogen-like molecules such as retinoids have been shown to have specific activity on sponge cells (Biesalski et al. 1992; Imsiecke et al. 1994), and they may negatively regulate telomerase activity, inducing differentiation of sponge cells, similar to their activity on cells of the hematopoietic system (Engelhardt et al. 1997). Finally, concomitant regulation of apoptosis and telomerase activity has been proposed (Mandal and Kumar 1997). Apoptosis has been demonstrated in sponges, and sponge cells might undergo terminal senescence by activation of central death signals, e.g. activation of the expression of the *MA-3* gene.

Acknowledgments. This work was supported by grants from the Bundesministerium für Bildung und Forschung (project "sponge cell culture") Germany and the International Human Frontier Science Program (RG-333/96-M).

References

Albanell J, Han W, Mellado B, Gunawardane R, Scher HI, Dmitrovsky E, Moore MA (1996) Telomerase activity is repressed during differentiation of maturation-sensitive but not resistant human tumor cell lines. Cancer Res 56:1503–1508

Arndt W (1930) Schwämme (Porifera, Spongien). Tabulae Biol 4(Suppl II):39–120

Arndt W (1935) Porifera. In: Grimpe G (ed) Tierwelt der Nord- und Ostsee. Akademische Verlagsgesellschaft, Leipzig, pp IIIa1–IIIa139

Ausubel FM, Brent R, Kingston RE, Moore DD, Smith JA, Seidmann JG, Struhl K (1995) Current protocols in molecular biology. Wiley, New York

Ax P (1995) Das System der Metazoa I. Fischer, Stuttgart

Bansal N, Houle A, Melnykovych G (1991) Apoptosis: mode of cell death induced in T cell leukemia lines by dexamethasone and other agents. FASEB J 5:211–216

Barnes RD (1987) Invertebrate zoology, 5th edn. Saunders, Philadelphia

Barthel D, Detmer A (1990) The spermatogenesis of *Halichondria panicea* (Porifera, Demospongiae). Zoomorphology 110:9–15

Batel R, Bihari N, Rinkevich B, Dapper J, Schäcke H, Schröder HC, Müller WEG (1993) Modulation of organotin-induced apoptosis by the water pollutant methyl mercury in a human lymphoblastoid tumor cell line and a marine sponge. Mar Ecol 93:245–251

Bergquist PL (1978) Sponges. Hutchinson, London

Bestilny LJ, Brown CB, Miura Y, Robertson LD, Riabowol KT (1996) Selective inhibition of telomerase activity during terminal differentiation of immortal cell lines. Cancer Res 56:3796–3802

Biesalski HK, Doepner G, Tzimas G, Gamulin V, Schröder HC, Batel R, Nau H, Müller WEG (1992) Modulation of *myb* gene expression in sponges by retinoic acid. Oncogene 7:1765–1774

Bode HR (1996) The interstitial cell lineage of hydra: a stem cell system that arose early in evolution. J Cell Sci 109:1155–1164

Borojevic R (1966) Étude experimentale de la différenciation des cellules de l'éponge au cours de son développement. Dev Biol 14:130–153

Borojevic R (1969) Étude du développement et la différenciation cellulaire d'éponges calcaires calcinéennes (genres *Clathrina* et *Ascandra*). Ann Embryol Morphol 2:15–36

Borojevic R (1970) Differenciation cellulaire dans l'embryogenèse et la morphogenèse chez les éponges. Symp Zool Soc Lond 25:467–490

Bowen ID, Bowen SM (1990) Programmed cell death in tumours and tissues. Chapman and Hill, New York

Buttke TM, Sandstrom PA (1994) Oxidative stress as a mediator of apoptosis. Immunol Today 15:7–10

Carroll SB (1995) Homeotic genes and the evolution of arthropods and chordates. Nature 376:479–485

Celerin M, Ray JM, Schisler NJ, Day AW, Stetler-Stevenson WG, Laudenbach DE (1996) Fungal fimbriae are composed of collagen. EMBO J 15:4445–4453

Chiu CP, Dragowska W, Kim NW, Vaziri H, Yui J, Thomas TE, Harley CB, Landsdorp PM (1996) Differential expression of telomerase activity in hematopoietic progenitors from adult human bone marrow. Stem Cell 14:239–248

Clark RB (1997) Marine pollution. Clarendon Press, Oxford

Counter CM, Avilion AA, LeFeuvre CE, Stewart NG, Greider CW, Harley CB, Bacchetti S (1992) Telomere shortening associated with chromosome instability is arrested in immortal cells which express telomerase activity. EMBO J 11:1921–1929

Davidson EH (1989) Lineage-specific gene expression and its regulative capacities of the sea urchin embryo: a proposed mechanism. Development 105:421–445

Davidson EH, Peterson KJ, Cameron RA (1995) Origin of bilaterian body plans: evolution of developmental regulatory mechanisms. Science 270:1319–1325

Degnan BM, Degnan SM, Giusti A, Morse DE (1995) A *hox/hom* homeobox gene in sponges. Gene 155:175–177

Denis H, Lacroix JC (1993) The dichotomy between germ line and somatic line, and the origin of cell mortality. Trends Genet 9:7–11

Diaz JP (1973) Cycle sexuel de deux demosponges de l'étang de Thau: *Suberites massa* Nardo et *Hymeniacidon sanguinea* Bowerbank. Bull Soc Zool Fr 98:145–156

Diaz JP (1979) Variations, différentiations et fonctions des catégories cellulaire de la demosponge d'eau saumaitres, *Suberites massa* Nardo, au cours du cycle biologique annuel et dans des conditions expérimentales. PhD Thesis, University of the Langedoc, Montpellier, pp 1–332

Egami N, Ishii S (1956) Differentiation of sex cells in united heterosexual halves of the sponge *Tethya serica*. Ann Zool Jpn 29:199–201

Engelhardt M, Kumar R, Albanell J, Pettengel R, Han W, Moore MA (1997) Telomerase regulation, cell cycle, and telomere stability in primitive hematopoietic cells. Blood 90:182–193

Gilbert W (1978) Why genes in pieces? Nature 271:501

Goldstein S (1990) Replicative senescence: the human fibroblast comes of age. Science 249:1129–1133

Gramzow M, Schröder HC, Fritsche U, Kurelec B, Robitzki A, Zimmermann H, Friese K, Kreuter MH, Müller WEG (1989) Role of phospholipase A2 in the stimulation of sponge cell proliferation by the homologous lectin. Cell 59:939–948

Hallmann A, Rappel A, Sumper M (1997) Gene expression by homologous recombination in the multicellular green algae *Volvox carteri*. Proc Natl Acad Sci USA 94:7469–7474

Harley CB (1991) Telomere loss: mitotic clock or genetic time bomb? Mutat Res 256:271–282

Harley CB (1995) Telomeres in aging. In: Blackburn EH, Greider CW (eds) Telomeres. Cold Spring Harbor Laboratory Press, Cold Spring Harbor, pp 247–263

Harley CB, Futcher AB, Greider CW (1990) Telomeres shorten during ageing of human fibroblasts. Nature 345:458–460

Harley CB, Kim NW, Prowse SI, Weinrich KS, Hirsch MD, West MD, Bacchetti S, Hirte HW, Counter CM, Greider CW, Wright WE, Shay JW (1994) Telomerase, cell mortality, and cancer. Cold Spring Harbor Symp Quant Biol 59:307–315

Hastie ND, Dempster M, Dunlop AG, Thompson AM, Green DK, Allshire RC (1990) Telomere reduction in human colorectal carcinoma and with ageing. Nature 346:866–868

Higashijima S, Kojima T, Michiue T, Ishimaru S, Emori Y, Saigo K (1992) Dual Bar homeo box genes of *Drosophila* required in two photoreceptor cells, R1 and R6, and primary pigment cells for normal eye development. Genes Dev 6:50–60

Holland PP, Ingham S, Krauss P (1988) Development and evolution. Mice and flies head to head. Nature 358:687–690

Holt SE, Wright WE, Shay JW (1996) Regulation of telomerase activity in immortal cell lines. Mol Cell Biol 16:2932–2939

Hyman LH (1940) The invertebrates; Protozoa through Ctenophora. McGraw-Hill, New York

Imsiecke G, Borojevic R, Custodio M, Müller WEG (1994) Retinoic acid acts as a morphogen in freshwater sponges. Invertebr Reprod Dev 26:89–98

Kay HR (1990) Reproduction in West Indian commercial sponges: oogenesis, larval development, and behavior. In: Rützler K (ed) New perspecitives in sponge biology. Smithsonian Institution Press, Washington, pp 161–169

Kim NW, Piatyszek MA, Prowse KR, Harley CB, West MD, Ho PLC, Coviello GM, Wright WE, Weinrich SL, Shay JW (1994) Specific association of human telomerase activity with immortal cells and cancer. Science 266:2011–2014

Kirkwood TBL, Rose MR (1991) Evolution of senescence: late survival sacrificed for reproduction. Philos Trans R Soc Lond [Biol] 332:15–24

Koziol C, Borojevic R, Steffen R, Müller WEG (1998) Sponges (Porifera) model systems to study the shift from immortal to senescent somatic cells: the telomerase activity in somatic cells. Mech Ageing Dev 100:107–120

Kruse M, Gamulin V, Cetkovic H, Pancer Z, Müller IM, Müller WEG (1996) Molecular evolution of the metazoan kinase C multigene family. J Mol Evol 43:374–383

Lévi C (1956) Étude des *Halisarca* de Roscoff embryologie et systematiques des demosponges. Arch Zool Exp Gen 93:1–181

Li WH, Graur D (1991) Fundamentals in molecular evolution. Sinnauer, Sunderland, Massachusetts

Liaci L, Sciscioli M (1975) Modalita di riproduzione sessuale di alcune poecilosclerina (Porifera). Atti Soc Peloritana Sci Fis Mat Nat 21:109–114

Liaci L, Sciscioli M, Matarrese A, Giove C (1971) Osservazioni sui sessuali di alcune keratosa (Porifera) e loro interesse megli studi filogenetici. Atti Soc Pelorit Sci Fisic Matem Natur 17:33–52

Lipps JH, Signor PW (eds) (1992) Origin and early evolution of Metazoa. Plenum Press, New York

Livingstone DR, Arnold R, Chipman K, Kirchin M (1989) The mixed function oxygenase system in molluscs: metabolism, response to xenobiotics and toxicity. Biochem Mar Organ 16:331–347

Mandal M, Kumar R (1997) Bcl-2 modulates telomerase activity. J Biol Chem 272:14183–14187

Margulis L, Schwartz KV (1995) Five kingdoms: an illustrated guide to the phyla of life on earth. Freeman, New York

McGinnis W, Krumlauf R (1992) Homeobox genes and axial patterning. Cell 68:283–302

Mehlhorn H (ed) (1989) Grundriß der Zoologie. Fischer, Stuttgart

Morin GB (1989) The human telomere terminal transferase enzyme is a ribonucleoprotein that synthesizes TTAGGG repeats. Cell 59:521–529

Müller WEG (1995) Molecular phylogeny of Metazoa (animals): monophyletic origin. Naturwissenschaften 82:321–329

Müller WEG (1997a) Origin of metazoan adhesion molecules and adhesion receptors as deduced from their cDNA analyses from the marine sponge *Geodia cydonium*. Cell Tissue Res 289:383–395

Müller WEG (1997b) Evolution of Protozoa to Metazoa. Theory Biosci 116:145–168

Müller WEG (1998a) Molecular phylogeny of Eumetazoa: experimental evidence for monophyly of animals based on genes in sponges (Porifera). Prog Mol Subcell Biol 19:89–132

Müller WEG (1998b) Origin of Metazoa: sponges as living fossils. Naturwissenschaften 85:1–15

Müller WEG, Müller IM (1998) Transition from Protozoa to Metazoa: an experimental approach. Prog Mol Subcell Biol 19:1–22

Müller WEG, Schröder HC, Ushijima H, Dapper J, Bormann J (1992) Gp120 of HIV-1 induces apoptosis in rat cortical cell cultures: prevention by memantine. Eur J Pharmacol 226:209–214

Müller WEG, Müller IM, Gamulin V (1994) On the monophyletic evolution of the Metazoa Brazil. J Med Biol Res 27:2083–2096

Müller WEG, Müller IM, Rinkevich B, Gamulin V (1995) Molecular evolution: evidence for the monophyletic origin of multicellular animals. Naturwissenschaften 82:36–38

Mutwakil MHAZ, Reader JP, Holdich DM, Smithurst PR, Candido EPM, Jones D, Stringham EG, De Pomerai DI (1997) Use of stress-inducible transgenic nematodes as biomarkers of heavy metal pollution in water samples from an English river system. Arch Environ Contam Toxicol 32:146–153

Nielsen C (1995) Animal evolution. Oxford University Press, Oxford

Nikoh N, Iwabe N, Kuma K, Ohno M, Sugiyama T, Watanabe Y, Yasui K, Schi-cui Z, Hori K, Shimura Y, Miyata T (1997) An estimate of divergence time of Parazoa and Eumetazoa and that of Cephalochordata and Vertebrata by aldolase and triose phosphate clocks. J Mol Evol 45:97–106

Norrback KF, Dahlenborg K, Carlsson R, Roos G (1996) Telomerase activity in normal B lymphocytes and non-Hodgkin's lymphomas. Blood 88:222–229

Ohno S (1970) Evolution by gene duplication. Springer, Berlin Heidelberg New York

Oncor (1996) TRAPeze telomerase detection kit, catalog no S7700-Kit, 2nd edn. Oncor, Gaithersburg, MD

Ottilie S, Raulf F, Bernekow A, Hanning G, Schartl M (1992) Multiple *src*-related kinase genes, *srk-1-4*, in the fresh water sponge *Spongilla lacustris*. Oncogene 7:1625–1630

Pahler S, Krasko A, Schütze J, Müller IM, Müller WEG (1998) Isolation and characterization of a cDNA encoding a potential morphogen from the marine sponge *Geodia cydonium*, that is conserved in higher metazoans. Royal Society, London (in press)

Patthy L (1995) Protein evolution by exon-shuffling. Springer Berlin Heidelberg New York

Paulus W (1989) Ultrastructural investigation of spermatogenesis in *Spongilla lacustris* and *Ephydatia fluviatilis* (Porifera, Spoingillidae). Zoomorphology 109:123–130

Philippe H, Chenuil A, Adoutte A (1994) Can the Cambrian explosion be inferred through molecular phylogeny? Development 120(Suppl):15–25

Ramirez RD, Wright WE, Shay JW, Taylor RS (1997) Telomerase activity concentrates in the mitotically active segments of human follicles. J Invest Dermatol 108:113–117

Rasmont R (1963) Le rôle de la taille et de la nutrition dans le déterminisme de la gemmulation chez les spongillides. Dev Biol 8:243–271

Reitner J, Mehl D (1996) Monophyly of the Porifera. Verh Naturwiss Ver Hamburg 36(NF):5–32

Richelle-Maurer E, Van de Vyver G, Visser S, Coutinho CC (1998) Homeobox-containing genes in freshwater sponges: characterization, expression, and phylogeny. Prog Mol Subcell Biol 19:157–175

Sarà M (1961) Richerche sul gonocorismo ed ermafroditismo nei Poriferi. Boll Zool 28:47–59

Schäcke H, Schröder HC, Gamulin V, Rinkevich B, Müller IM, Müller WEG (1994) Molecular cloning of a tyrosine kinase gene from marine sponge *Geodia cydonium*: a new member belonging to the receptor tyrosine kinase class II family. Mol Membr Biol 11:101–107

Seimiya M, Naito M, Watanabe Y, Kurosawa Y (1998) Homeobox genes in the freshwater sponge *Ephydatia fluviatilis*. Prog Mol Subcell Biol 19:133–155

Shibahara K, Asano M, Ishida Y, Aoki T, Koike T, Honjo T (1995) Isolation of a novel gene *MA-3* that is induced upon programmed cell death. Gene 166:297–301

Simpson TL (1984) The cell biology of sponges. Springer, Berlin Heidelberg New York

Slack JMW (1987) Morphogenetic gradients – past and present. Trends Biochem Sci 12:200–204

Stanley PE, Kricka LJ (1990) Bioluminescence and chemiluminescence; current status. Wiley, New York

Strehler BL (1986) Genetic instability as the primary cause of human aging. Exp Gerontol 21:283–319

Thompson CB (1995) Apoptosis in the pathogenesis and treatment of disease. Science 267:1456–1462

Valentine JW (1994) Late Precambrian bilaterians: grades and clades. Proc Natl Acad Sci USA 91:6751–6757

Van de Vyver G, Willenz P (1975) An experimental study of the life cycle of the fresh-water sponge *Ephydatia fluviatilis* in its natural surroundings. Wilhelm Roux's Arch Dev Biol 177:41–52

Van Soest RWM (1994) Demosponge distribution pattern. In: Van Soest RWM, van Kempen TMG, Braekman JC (eds) Sponges in time and space. Balkema, Rotterdam, pp 213–223

Vojta PJ, Barrett JC (1995) Genetic analysis of cellular senescence. Biochim Biophys Acta 1242:29–41

Weismann A (1892) Über die Vererbung. Fischer, Jena

Willmer PG (1994) Invertebrate relationships. Cambridge University Press, Cambridge

Wolpert L (1990) The evolution of the development. Biol J Linn Soc 39:109–124

Wyllie AH (1980) Glucocorticoid-induced thymocyte apoptosis is associated with endogenous nuclease activation. Nature 284:555–556

Yasumoto S, Kunimura C, Kikuchi K, Tahara H, Ohji H, Yamamoto H, Ide T, Utakoji T (1996) Telomerase activity in normal human epithelial cells. Oncogene 13:433–439

Zahn RK, Müller I, Müller WEG (1977) *Amoibodictya forsteri* n. g. n. sp. und die Untergliederung der Esperiopsidae (Porifera: Poecilosclerida). Senckenb Biol 58:105–112

The Notion of the Cambrian Pananimalia Genome and a Genomic Difference that Separated Vertebrates from Invertebrates

S. Ohno[1]

1
Introduction

When our knowledge of early Cambrian fossils was almost entirely derived from the Burgess fauna in British Colombia, Canada, the so-called Cambrian explosion that started roughly 540 million years ago appeared to have been a rather slow detonation lasting 20 to 30 million years (Conway Morris 1989). The subsequent discovery of other early Cambrian faunas such as the Chenjiang fauna in Yunnan province, China, forced us to reevaluate the above noted initial estimate. It now appears that the Cambrian explosion was indeed a literal explosion, ancestral forms belonging to nearly all the extant animal phyla came into being within the short period of 6 to 10 million years (Gould 1995). In the case of the phylum Chordata, to which our own species belongs, its three tiers, namely Urochordata, Cephalochordata and Vertebrata, too appear to have emerged almost simultaneously within this short duration.

Unless various prior events that led to the Cambrian explosion are understood, no significant insight can be gained into the Cambrian explosion itself. Accordingly, this chapter starts from the beginning of life on this earth.

2
Cyanobacteria in the Archean Ocean

Our own solar system is thought to have come into being some 4.7 billion years ago. During the first billion years, however, this earth was constantly bombarded by meteorites to the extent that the total energy released was enough to evaporate the entire sea water of the lifeless primeval ocean (Chang 1994). It would appear that the primeval ocean of this earth during the first one billion years of its existence had been totally incompatible either with life in any form or with various reactions of the prebiological world (e.g., RNA world) that led to the creation of life on this earth.

Therefore, it is with great surprise that one finds evidence of flourishing microbes in the early Archean ocean of 3.0 to 3.5 billion years ago. On the other

[1]Beckman Research Institute Df the City of Hope, 1450 East Duarte Road, Duarte, California 91010, USA

Progress in Molecular and Subcellular Biology, Vol. 21
W.E.G. Müller (Ed.)
© Springer-Verlag Berlin Heidelberg 1998

hand, there is no surprise in the fact that these earliest microbial fossils in the Archean apex chert from Australia were primarily cyanobacteria (Schopf 1993).

Inasmuch as heavier elements of this universe were and are being created from hydrogen in the inferno of nuclear fusion reactions that are supernova, all of the molecular oxygen just born is consumed in the act of burning to combine with other elements. Accordingly, H_2O and CO_2 are two of the most abundant molecules in drifting molecular clouds of this universe. It follows that while the early Archean ocean was saturated with CO_2, the main ingredient being $NaHCO_3$, it was devoid of molecular oxygen (Kempe and Degens 1985). It stands to reason that the earliest living forms on this earth were photosynthesizing cyanobacteria that were capable of utilizing carbon dioxide in the anaerobic environment. It was continuous release of molecular oxygen by these photosynthesizing cyanobacteria that started to set the stage for the eventual Cambrian explosion that almost simultaneously yielded all the diverse animal phyla.

There was yet another great contribution made by these early cyanobacteria in that they transferred their photosynthesizing capability to a particular eukaryote by becoming an endosymbiont that eventually transformed itself to one type of cytoplasmic organelle peculiar to plants, known as chloroplasts (Cavalier-Smith 1987). The plant form compatible with early anaerobic aquatic life is a special form of the blue-green algae; a plant form with chloroplasts but without mitochondria. Unfortunately, such a blue-green alga is not known.

3
Archaezoa as the First Animal in the Early Anaerobic Environment?

Generally speaking, photosynthesizing plants are consumers of carbon dioxide and releasers of molecular oxygen, whereas animals are consumers of molecular oxygen and releasers of carbon dioxide. If one follows this common definition, the first animal on this earth should have appeared only after the molecular oxygen concentration in the ocean reached a level that could be considered as aerobic. Such an aerobic condition could have been achieved only after the long toil of free oxygen release by photosynthesizing cyanobacteria, which were much later probably joined by early blue-green algae. Yet today there are unicellular eukaryotic animals such as parasitic trichomonads that can thrive in an anaerobic environment. The very fact that they are classified as archaezoa indicates that their ancestral forms might have come into being soon after the emergence of the first living creature on this earth in the form of cyanobacteria 3.5 billion years ago; say within a billion years after the emergence of cyanobacteria. The archaezoan acquired the ability to live under anaerobic conditions from an anaerobic bacterium that became an endosymbiont and later transformed itself into a hydrogenosome; yet another cytoplasmic organelle derived from a parasitic bacterium. Although its prokaryotic circular genome is no longer found inside the hydrogenosome, the presence of genes in the nuclear genome of trichomonads that encode the prokaryote type of heat shock proteins testifies to the past transfer of pertinent genes from the hydrogenosome genome to the nucleus (Bui et al. 1996). Inside the hydrogenosome, ATP is generated without help from free oxygen, by succinyl coenzyme A, mainly utilizing pyruvate as a substrate. In the process, hydrogenosomes release molecular hydrogen; hence the name.

4
The Acquisition of Mitochondria, Derived from an Endosymbiotic Paracoccus-Like Purple, Nonsulfur Bacterium as a Conditio Sine Qua Non to the Cambrian Explosion

Since the Cambrian explosion that started roughly 540 million years ago involved only aquatic animals of the ocean and not plants, the trigger for this explosion had to be the aerobic condition of the ocean, finally provided by the 3.5-billion-year-long diligence of photosynthesizing cyanobacteria. It follows that pre-Cambrian animals had to make themselves ready for the coming explosive diversification by acquiring the ability to live under aerobic conditions. This they did by choosing one aerobic, purple, nonsulfur bacterium as an endosymbiont, one billion years ago (de Duve 1991). A living relative of this endosymbiont has been identified as *Paracoccus denitroficans* dubbed "a free-living mitochondrion" (van Verseveld and Bosma 1987). Genes for the set of enzymes of the citric acid cycle that form the heart of oxidative carbohydrate metabolism to generate ATP have long been transferred from the mitochondrial genome to the nucleus, and so were nearly 700 other pertinent genes. Yet the mitochondrion of today still retains its own circular genome, even though very small, containing genes for prokaryotic *ribosomal* RNAs and for the set of its own *transfer* RNAs as well as a small number of protein-coding genes; e.g., cytochrome b gene. In view of the fact that continuous transfer of mitochondrial genes to the nuclear genome are still occurring even in our own species (Collura and Stewart 1995, Zischler et al. 1995), the retention of mitochondrial genomes of variable sizes by nearly all eukaryotes begs a rational explanation. The answer appears to be found in the fact that mitochondrial genomes, by retaining their own set of 22 *transfer* RNA genes, managed to develop their own genetic codes that are distinct from the universal genetic code still in operation in the nuclear genome. Accordingly, any mitochondrial gene of today becomes a useless pseudogene when transferred to the nucleus. For example in the panmetazoan mitochondrial code, UGA is a Trp codon; AGA and AGG are Ser condons in the invertebrate mitochodrial code, while they function as stop condons in the vertebrate mitochodrial code. In the nuclear universal code, however, UGA stands for stop codon, whereas AGA and AGG are Arg codons. At the beginning of endosymbiosis 1 billion years ago, both a parasitic bacterium as well as its eukaryotic host must have operated on the universal code, thus allowing free transfer of prococcal genes to the nuclear genome. When the mitochondrion finally chose to establish its own genetic code, this one-way transfer became nonsensical, thus allowing mitochondria to retain their own genome even though much reduced in size.

Since it appears that successful symbiosis between a parasitic bacterium and a eukaryotic host was feasible only when the latter was still unicellular, it follows that the common ancestor of all the Cambrian animals that represent nearly all extant animal phyla was unicellular one billion years ago. Put it in another way, the ancestral animal was still unicellular 460 million years prior to the Cambrian explosion. Needless to say, this unicellular eukaryote of 1 billion years ago was also likely to have been ancestral to all other eukaryotes with mitochondria that would have included yeasts, plants and others. If it is further shown that the same set of genes is retained by mitochondrial genomes of all the diverse metazoan animal phyla, this would lead us to only one conclusion; that a few hundred million years prior to the Cambrian explosion, there was still only one animal form that served as the common ancestor of all the divergent Cambrian animal forms. This ancestral animal form that existed roughly 740 million years ago might have already been multicellular. But again, it might have still been unicellular. Indeed, it appears that the universal formula of two genes for 12S and 16S *ribosomal* RNAs, the complete set of 22 genes for *transfer* RNAs and 13 additional genes encoding different polypeptides applies to mitochondrial genomes of nearly all the metazoan animal phyla that emerged in the Cambrian explosion. As shown in Table 1, these 13 genes encode seven subunits of NADH dehydrogenase, two subunits of ionized hydrogen ATPase, three subunits of cytochrome oxidase and finally cytochrome b itself. The genetic content of mitochondria derived from phylum Chordata (represented by human, chicken, frog and various fish) are identical with that of the phylum Echinodermata represented by a sea urchin (Jacobs et al. 1988), as well as the phylum Arthropoda symbolized by a fruit fly. Most remarkable is the fact that with regard to three diverse phyla mentioned above, the coding sequences of 36 of the 37 mitochondrial genes reside on the same DNA strand, whereas a remaining solitary gene encoding only a 173-residue-long ND6 subunit is invariably on its complementary strand. There remains little doubt that members of the above three diverse metazoan phyla were still represented by a single common ancestor barely a few hundred million years before the Cambrian explosion.

Problematical is the finding from the mitochondrial genome of a parasitic nematode, *Ascaris suum*, representing the phylum Nemathelminthes (Wolstenholme et al. 1987). Here, it was reported that a gene for the ATPase 8 subunit is missing from its mitochondrial genome, and an ND6 subunit gene resides on the same DNA strand as the rest of mitochondrial genes (Table 1). Unless the above is a quirk brought about by a parasitic way of life, one has to conclude that a few hundred million years before the Cambrian explosion, there was not a single common ancestor, but a small number of closely related

Table 1. The identical set of 37 genes retained by the mitochondrial genome of the three metazoan phyla (left column) is contrasted with a slightly different set possessed by the mitochondrial genome of a parasitic round worm of another metazoan order; Nemathelminthes (middle column). Shown in the right column is a different set of mitochondrial genes possessed by baker's yeast, representing the kingdom Fungi

Genes in the mitochondrial genome

Chordata (human, chicken, frog, fish) Echinodermata (sea urchin) Arthropoda (fruit fly)	Nemanthelminthes (*Ascaris* roundworm)	Outgroup Baker's yeast
2 Large *ribosomal* RNAs	Same	Same
22 *Transfer* RNAs	Same	Same
Missing	Missing	1 Small *ribosomal* RNA
NADH Dehydrogenase subunits		
ND1	Same	Missing
ND2	Same	Missing
ND3	Same	Missing
ND4	Same	Missing
ND4L	Same	Missing
ND5	Same	Missing
ND6	ND6	Missing
H$^+$ ATPase Subunits		
AT6	Same	Same
AT8	Missing	AT9
Missing	Missing	Same
Cytochrome oxidase subunits		
COI	Same	Same
COII	Same	Same
COIII	Same	Same
Cytochrome b		
Cytb	Same	Same
Missing	Missing	5 Nucleic acid specific enzymes

ancestors that gave rise to diverse metazoan animal phyla at the time of explosive diversification. In Table 1, the mitochondrial genetic content of baker's yeast (*Saccharomyces cerevisiae*) is also used as an outgroup to show the evolutionary significance of the near identity in genetic contents of metazoan mitochondria (Attardi 1988). As far as the larger picture is concerned, it appears that when mitochondrial genomes succeeded in preventing further erosion of their integrities by evolving their own unique genetic codes, their hosts, as members of the superkingdom Metakaryota, had already diverged to form at least three kingdoms; Animalia, Fungi and Plantae.

5
Ediacaran Emergence of Porifera and Cnidaria as a Prelude to the Cambrian Explosion

During the 20 million years or so before the beginning of the Cambrian period, before the Paleozoic era, animals belonging to two phyla made their appearances. This period came to be known as the Ediacaran episode (Cloud and Glaesner 1982). Animals represented in this pre-Paleozoic episode were sponge-like creatures apparently belonging to the phylum Porifera and jellyfish-like creatures with an obvious affinity to the phylum Cnidaria. It is of interest that members of these two phyla were only diploblastic, having ectoderm and endoderm but still lacking a mesoderm. Consequently, there was no secondary body cavity known as coelom, instead a gelatinous layer of varying thickness called the mesogloea occupied the space between the ectoderm and the endoderm. However, not all the ediacaran animals were of the simpler body design, for there were others showing affinities to other presumably more advanced phyla such as Annelida and Mollusca. By far the most bizarre was a round but oblong flat worm made of about 60 radially arranged body segments named *Dickinsonia* that apparently showed an affinity toward Annelida. Furthermore, this *Dickensonia* did not perish but persisted into the early Cambrian.

Here, it is of importance to point out that the pre-Cambrian ocean in which the creatures noted above began to appear became very different from the Archaen ocean. It, in fact, came to resemble the ocean of today. It should be recalled that the Archean ocean, containing $NaHCO_3$ as a main ingredient, was alkaline (pH 9.5–10.5), and this forced the calcium concentration in the Archean ocean to remain at the level of 10^{-7}M. Incidentally, this calcium concentration became a permanent fixture as the universal intracellular calcium concentration in all the animals. Any calcium in excess of the above became insoluble precipitates such as calcite, aragonite or dolomite (Kempe and Degens 1985). Subsequently, "titration of the ocean by acid volatiles such as HCl" continued, and as a result, NaCl replaced $NaHCO_3$ as the main ingredient in the pre-Cambrian ocean, thereby lowering the CO_2 concentration to only one-thousandth of the Archean level, while its calcium concentration went up 100-fold or more and, of course, previously nonexistent molecular oxygen now became abundant (Degens et al. 1984).

6
Animals of the Cambrian Explosion and the Simultaneous Emergence of Three Subphyla of the Phylum Chordata

Multitudes of animals that burst forth in the Cambrian explosion included varieties of worms showing affinities to such phyla as Nemertina, Annelida, Nemathelminthes and others. The phylum Mollusca was represented by at least two separate classes; shellfish of Bivalvia and squid-like creatures of Cephalopoda. Since only marine animals were involved in this explosion, a large variety of crustaceans represented the phylum Arthropoda. One crustacean-like creature, *Opabinia regalis*, had five eyes and a mouth with a pair of scissor-like tongs attached to the end of a long tube, while another creature of considerable size, *Anomalocaris*, is said to have been the greatest predator of the early Cambrian sea. However, among the early Cambrian arthropods, there were also trilobites that were to become numerous, only to perish at the end of the Paleozoic era 200 million years ago. The phylum Echinodermata was represented by sand dollar-like creatures. Needless to say, there were a number of truely bizarre animals that defied classification such as *Hallucigenia*, with six pairs of stilt-like legs on the ventral as well as the dorsal side. From two opposite sides of the anal opening of another worm-like creature protruded a pair of well calcified shells, as though a bivalve had parasitized this area of a worm (Conway Morris 1989).

As to the emergence of vertebrates within the phylum Chordata, the scenario widely accepted until 10 years ago was as follows: (1) Early in the Cambrian, there emerged a group of sessile filter feeders constituting the subphyla Hemichordata and Urochordata. Although the adult body consisted of an encased digestive tract and little else, in tadpole-like larvae of the latter subphylum, the basic vertebrate body design emerged. (2) When, by the process of paedomorphosis, these larvae acquired reproductive capacity, thereby abolishing the adult stage, amphioxus-like creatures of the subphylum Cephalochordata emerged. (3) By developmentally replacing the stout notochord with a series of vertebral bones, the first vertebrate in the form of agnatan (jawless) fish evolved in the Ordovician period roughly 100 million years after the Cambrian explosion.

During the last few years, the seemingly very sensible scenario described above was shattered by the following discoveries. Early Cambrian fossils of the Chenjiang fauna from Yunnan province of China contained those of *Yunnanozoon lividum*, unmistakably an amphioxus-like creature representing the subphylum Cephalochordata (Chen et al. 1995). Sets of bony plates bearing delicate file-like teeth have long been a permanent fixture among the Cambrian fossils of various ages, including the earliest ones, and, for this reason, the unknown host of such a set, consisting of about five pairs of bony plates bearing file-like teeth came to be defined as the conodont. This conodont turned out to be a very large eyed agnathan (jawless) fish (Gabbot et al. 1995).

This recent revelation was remarkable insofar that instead of the most ancestral subphylum Urochordata evolving into the next in the hierarchical order, the Cephalochordata and finally to the last subphylum Vertebrata in steady progression, as one would have expected, all the three tiers of the phylum Chordata in fact made an almost simultaneous appearance, as if to prove that these three subphyla merely represented three alternative manifestations of the same genetic resource.

7
Genes in the Cambrian Pananimalia Genome

In view of the above, I have proposed that all the varieties of early Cambrian animals that made their almost simultaneous appearances must necessarily have been endowed with nearly the identical genome which was defined as the **Cambrian Pananimalia genome** (Ohno 1996). Indeed, the remaining genes in various mitochondrial genomes listed in Table 1 clearly indicate that those divergent early Cambrian animals shared a very small number of closely related ancestors, if not a single common ancestor, only a few hundred million years before the explosion.

Not surprisingly, the same gene, time and again, was found to be carrying out not only obviously the same task but also apparently different tasks in metazoa of diverse phyla, thus enhancing the credibility of a notion of the pananimalia genome. For example, if one focuses on the phylum Chordata alone, one would have easily been misled to believe that a gene or genes for hemoglobins as intercellular oxygen carriers, arose de novo in the subphylum Vertebrata, since hemoglobin molecules are not found in either Urochordata or Cephalochordata. The fact is that a gene or genes encoding hemoglobins had to be included in the Cambrian pananimalia genome, for hemoglobins are found in individual species belonging to nearly all the animal phyla such as Nemertina, Annelida, Mollusca, Arthropoda and Echinodermata. While the above illustrates an instance of individual species exercising an option of use or disuse of particular pananimalia genes, others in the pananimalia genome were and are utilized by all metazoa. Genes encoding various forms of stoutly fibrous collagens are such examples, for without fibrous collagens to hold together external body parts and internal organs, there could have been no metazoan.

Inasmuch as early Cambrian animals displayed dazzling varieties of body forms, we shall now focus on pananimalia regulatory genes that encoded various DNA binding transcription factors, thereby shaping development of body parts.

7.1
Pax 6 Genes and Eye Formation

Pax genes are defined by the presence of a conserved paired box that encodes a 128-residue-long paired domain. These genes are mostly involved in development of the central nervous system. In the mouse, for example, Pax 2 and Pax 5 genes are involved in the shaping of the midbrain and hindbrain in a somewhat functionally redundant and overlapping manner; Pax2 $(+/-)$, Pax5 $(-/-)$ mice manifesting complete loss of the posterior midbrain and cerebellum. In addition, these two Pax genes in the mouse play auxillary roles that were probably secondary aquisitions; Pax 2 concerning itself with metanephros development, while Pax 5 involves itself in early differentiation of B lymphocytes (Urbanek et al. 1997). At any rate, the most remarkable so far among the Pax genes has been the Pax 6 gene which apparently directs eye development in animals of all the divergent phyla. In a situation quite analogous to that already mentioned with regard to the presence or absence of hemoglobins, eyes are possessed by members of the subphyla Vertebrata and Cephalochordata but not of Urochordata within the phylum Chordata. Eyes are already found among members of the genus *Hydra*, although this genus belongs to the Ediacaran phylum Cnidaria, and among flatworms of Platyhelminthes, planarians are endowed with eyes and so are ribbon worms of Nemertina. Within the phylum Mollusca, members of Cephalopoda and certain members of Gastropoda develop eyes, while those of Bivalvia do not. Yet, whenever developmental studies were made on animals with eyes, the local expression of Pax 6 gene, confined to the regions where eyes were to develop, has invariably been observed, regardless of the phylal affiliation (Loosli et al. 1996). It has been known for some time that *Drosophila* fails to develop eyes in the Pax6 ($-/-$) state. It has recently been shown that the rat in the Pax6 ($-/-$) state also fails to develop eyes. In addition, a nose was lost as well from Pax6 ($-/-$) rats (Matsuo et al. 1993). It should be stressed here that as far as eyes go, not only are the compound eyes of adult insects unusual, but so are vertebrate eyes, because of their inverted development, with photoreceptors on cones and rods of the retina facing the sclera instead of the lens. Yet all the eyes are apparently shaped under the direction of the Pax 6 gene. Of all the metazoa with eyes, only hydra belonged to the Ediacaran phylum, and the Ediacaran genome might have been a forerunner of the Cambrian pananimalia genome. Indeed, Pax 6 gene sensu stricto was not found in the hydra genome. Instead, there was Pax B gene that is equally related to Pax 4, Pax 6 as well as to Pax 2, Pax 5 and Pax 8 (Sun et al. 1997).

The very fact that Pax 6 ($-/-$) rats failed to develop a nose revealed that aside from eyes, this gene may also control shaping of olfactory and other sensory organs. Indeed, the Pax 6 gene is alive and well in the genome of such animals as eyeless sea urchins and equally eyeless namatodes.

7.2
The Universal Control of Anterior-To-Posterior Body Segment Differentiation by a Closely Linked Set of Hox Genes

Included in the body construction plan of all metazoa is an early development of anterior-to-posterior body segments. While it is true that a particular nematode, *Caenorhabditis elegans*, does not develop from body segments, such must represent a secondary loss peculiar to the order Rhabditoidea, since chain-mail worms belonging to the other nematode order Desmoscolecoidea contain more than a dozen silicate ring skeletons inside their bodies. Similarly, it is true that the radial symmetry representing a centrifugal type of body segmentation which characterizes the adult body of sea urchins, is also a secondary development which starts on one side of the body of their free swimming larvae. Furthermore, the adult body of sea cucumbers belonging to the other echinodermal class Holothuroidea demonstrates the usual bilateral sysmmetry. In all animals, the anterior-to-posterior differentiation of body segments is under the control of a set of closely linked Hox (homeotic box) genes, the number of Hox genes in a set differing between 4 and 13. Among arthropods, crustaceans have only four Hox genes in a set, while insects are endowed with eight in a set. Furthermore, this set of Hox genes was already in the Ediacaran genome (Carroll 1995).

In *Drosophila* with a set of eight Hox genes, a defective mutation affecting the sixth Hox gene Ubx in a set, transforms a dipteran *Drosophila* to a dragon-fly-like tetraptera by transforming a pair of halters to the second set of wings, hence, the gene name *bithorax* (Lewis 1978). As shall be discussed in detail, vertebrates are unique in having four sets of 13 Hox genes each. At any rate, in the mouse, defective mutations simultaneously affecting Hox 3 genes of A and D series cause the complete loss of the first cervical vertebra known as atlas (Condie and Capecchi 1994). In the case of Hox 11 genes of the mouse, simultaneous disfunction of two Hox 11 also of A and D series cause the simultaneous disappearance of forearm bones; radius and ulna (Davis et al. 1995).

With regard to the true pananimalia nature of these sets of closely linked Hox genes, there has been no better demonstration than the one below: the deformed phenotype manifested by *Drosophila* with defective Hox *Dfd* gene can be reverted to the normal phenotype by a normal human Hox B4 gene introduced by transfection to mutant *Drosophila* (McGinnis et al. 1990).

7.3
The Antiquity to Ftz-F1, COUP and Other Genes Encoding Nuclear Receptor Proteins

Historically, this group of rather large proteins, on average 1000 residues-long, were found as receptors of various steroid hormones during the 1960s in mammals and birds. While receptors for peptide hormones invariably reside on the cell surface, receptors for steroid hormones are found inside the cell nucleus. Upon binding to its specific ligand, which is a steroid hormone, an activated nuclear receptor binds to a specific set of DNA sequences in the genome, thus inducing enhanced transcriptions of a specific set of genes. In this sense, these nuclear receptors were one of the first transcription factors discovered among metazoa. However, it soon became apparent that this family of DNA binding proteins categorized as NR (nuclear receptors) has a large membership; some of them utilizing retinoic acid, vitamin D, ecdyson, pros-taglandin and even oligopeptidic thyroid hormones as their ligands. Further-more, within this family, there exists another substantial group of nuclear receptors toward which no known ligand showed a binding affinity and this third group came to be known as orphan nuclear receptors. As it turned out, some of these genes encoding certain orphan nuclear receptors perform primordial developmental functions in animals belonging to diverse phyla, therefore they were likely to be included in the Cambrian pananimalia genome.

The case in point is Ftz-F1 gene first discovered in *Drosophila*. Ftz is one of the solitary homeobox genes of *Drosophila* concerned with very early zygote segmentations. Ftz is an abbreviation of the Japanese words "Fusi Tara Zu" meaning "not enough number of segments"; indeed, in the functional absence of Ftz, the normal seven striped pattern of early *Drosophila* embryos fails to develop, resulting in irregular, less than seven striped patterns. Ftz-F1 is a gene encoding a transcription activator of Ftz gene. Accordingly, Fiz-F1 $(-/-)$ state manifests a defect almost identical with that manifested by Fiz $(-/-)$ state (Lavorga et al. 1991). Ftz-F1 $(-/-)$ state in vertebrates caused an equally severe developmental defect, even though very different in its target; Fiz-F1 $(-/-)$ mice are not viable because of the complete absence of adrenals and gonads (Luo et al. 1994).

Seminal phylogenic work has recently been performed on NR (nuclear receptor) genes of metazoa (Escriva et al. 1997). They have shown that not only two kinds of orphan NRs, Fiz-F1 and Coup, but also the retinod receptor, RXR, are already present in a sea anemone representing the Ediacaran phylum Cnidaria. There remains little doubt that these NR genes were included in the Cambrian Pananimalia genome. Of the phylum Arthropoda, members of the Cambrian class Crustacea, appeared to have still retained the same set of NR genes as found in a sea anemone. In addition, however, the gene for ecdysone-receptor (ECR) was already found in this class. It should be recalled that ecdysone is a steroid-related hormone involved in the induction of insect metamorphosis. As to members of the newer class Insecta, genes encoding 15 to 20 diverse NRs were found in the *Drosophila* genome. In the case of the subphylum Vertebrata, the number of genes encoding even more diverse NRs approaches 50 (Escriva et al. 1997). All in all, the concept of the Cambrian Pananimalia genome appears to be based upon a solid foundation.

8
Two Successive Rounds of Tetraploidization Events at the Beginning of Vertebrate Evolution and the Invariable Presence of Tetralogous Genes in the Vertebrate Genome

By now, there remains little doubt that the redundancy created by the mechanism of gene duplication has been the prime driving force of evolution (Ohno 1970). Essentially, there are three different means of achieving gene duplication. The first is a tandem duplication of a gene or a set of closely linked genes on the same chromosome, most often by an unequal crossing over during meiosis. This can occur at any place on a chromosome at any time in any species. The second is an insertion of a DNA copy of the transcript to a position on another chromosome via retroviral reverse transcriptase. Since such a transcript is devoid of the transcription initiation signal base sequence, an inserted copy of the gene usually becomes a functionless pseudogene, thus joining the ranks of junkDNA (Ohno 1972). Although a transposon may also insert a viable gene to another place on another chromosome, detailed comparison of the human genome with the mouse genome only revealed the conservation of linked segments between two widely divergent mammalian species. There is little, if any, evidence of individual genes being placed outside of the conserved chromosomal segments. Accordingly, the second mechanism appears to have played no significant role in vertebrate evolution. The third, which is most pertinent to vertebrate evolution is the simultaneous duplication of the entire genome by becoming tetraploid. This third mechanism, however, is incompatible with the well established genetic sex determining mechanism. Therefore, it could have occurred only at the very beginning of vertebrate evolution. Indeed, even today, many tetraploid species are found among various fishes. For example, within the teleost suborder Cyprinoidea, the common carp (*Cyprinus carpio*, 2n = 104) is a tetraploid species in relation to other carps such as the grass carp (*Ctenopharyngodon idellus*, 2n = 52). Inasmuch as 3.14 to 5.91% amino acid sequence differences were seen between two copies of somatotropins, gonadotropins and prolactins invariably present in the common carp, this species is thought to have become tetraploid about 15 million years ago (Ohno 1970). It was thought that the first vertebrate in the form of a Cambrian jawless conodont fish was already a tetraploid in relation to its contemporary cephalochordates such as *Yunnanozoon* already noted. Thus, the subphylum Vertebrata was already a distinct genomic entity even at the time of the Cambrian explosion some 540 million yeras ago. The second tetraploidy event was thought to have taken place when an agnathan (jawless) fish evolved into the gnathostomes (jawed entity) during the Silurian period some 330 million years ago (Holland et al. 1994).

It has been well established in plants that there are two kinds of tetraploidies. The first is autotetraploidy; the literal duplication of the entire genome within the species. The second is allotetraploidy; an otherwise sterile interspecific hybrid, regaining fertility by becoming a tetraploid. During meiosis of an allotetraploid, homologous chromosomes of one species can complete longitudinal paring among themselves and so can homologous chromosomes of another species. Accordingly, an interspecific hybrid which is sterile in the diploid state due to the formation of multivalents during meiosis becomes fertile by becoming an allotetraploid. Although the reasons shall not be given here, both of the tetraploidy events that made the verbebrate genome unique among the metazoan genomes appear to have been allotetraploidy (Spring 1997).

8.1
Four Sets of Hox Genes on Tetralogous Regions of Human Chromosomes 7p12, 17q11.2–12, 12q13 and 2q34

As aleady mentioned, a set of closely linked Hox genes that control the craniocaudal differentiation of body segments was already present in the Ediacaran genome. The outstanding feature of this set is that although the number of Hox genes within a set may vary between three and 13, members of every metazoan phylum with the single exception of the subphylum Vertebrata, has been endowed with a single set of Hox genes. This is true even of the two subphyla Urochodata and Cephalochrodata of the phylum Chordata, although the number of Hox genes in a set has increased to 10 in Cephalochordata. As to our own subphylum Vertebrata, two groups of cyclostomes are survivors that still represent the most primitive agnathan (jawless) state of vertebrate evolution. Of the above two groups of cyclostomes, lampreys have two sets of Hox genes, numbering 13 in each set, whereas more primitive hagfish, having a genome twice the size of the lamprey genome, are already endowed with four sets of Hox genes as did all the gnathostomes; from jawed fish to mammals (Holland et al. 1994). Here, an exception to the rule among the vertebrates should be noted. New tetraploid species found among the fish and amphibians such as the common carp already noted, should now be endowed with eight sets of Hox genes. These rare exceptions, however, shall be put aside for the sake of keeping a larger evolutionary picture in focus. In the case of the human genome, at any rate, Hox A, Hox B, Hox C and Hox D series reside on four different chromosomes; autosomes No. 7, 17, 12 and 2 at the specific locations noted in the heading. These four chromosomal regions constitute but one of the many quadruplets of homeologous regions in the human as well as mouse genome earlier identified by Lundin (1993); lengths of these homologous regions vary between 2 and 20 million base pairs (Lundin 1993). Inasmuch as these homeologous regions were remains of the two successive tetraploidization events that occurred in the remote past of modern vertebrates, one might as well call these homologous regions tetralogous regions. For example, in the vicinity of HoxA, HoxB, HoxC and HoxD series of Hox genes, there is indeed another set of tetralogous genes encoding four varieties of EGF-(epidermal growth factor)-receptor tyrosine kinases, which are known in humans and mice as EGFR, ERBB2, ERBB3 and ERBB4 (Spring 1997).

8.2
Inevitable Degeneration of Tetralogous Genes to Trilogues, Dilogues and Even to Monologues

A copy of the gene created by the mechanism of gene duplication becomes functionally redundant. Being redundant, it escapes from relentless surveillance by natural selection to maintain status quo and freely accumulate randomly sustained base substitutions. Two alternative fates wait in its future; either, by acquiring a new function, it emerges triumphant as a new gene, or by failing in this high stakes gamble, it becomes a functionless pseudogene, thus joining the ranks of abundant junkDNA (Ohno 1972).

The first eukaryotic genome project to be completed was on baker's yeast (*Saccharomyces cerevisiae*) with 5800 genes in its genome. Since this genome still retains 55 duplicated regions, it was concluded that baker's yeast too is an ancient tetraploid species, this event taking place roughly 150 million years ago. However, by now only 13% of the total genes still retained their duplicate copies (Wolfe and Shields 1997). As expected from the randomness of these base substitutions, degeneracy is the more likely of the two alternative fates to be experienced by a redundant copy of the gene.

In the case of metazoa, however, it appears that these redundant copies retain their functions longer than those of baker's yeast, and they also enjoy a considerably greater chance of becoming a new gene with a hitherto nonexistent function. It should be recalled that Hox A, Hox B, Hox C, and Hox D series with 13 gene loci in each were born when the vertebrate genome underwent the second tetraploidy event roughly 330 million years ago (Holland et al. 1994). Both in the mouse and human genomes, the tetralogous situation is still maintained by Hox 4th and Hox 9th loci, with all four genes (A4, B4, C4 and D4) still functioning. In the case of Hox 2nd, Hox 7th and Hox 12th loci, by contrast, tetralogues degenerated into dilogues; only two of the four genes (C12 and D12) remaining functional. As to the rest, the trilogous situation prevailed in eight Hox loci. In short, 13 of the 54 Hox genes (24%) joined the rank of junkDNA after 330 million years. This is considerably less than the attrition rate observed among the duplicated loci of baker's yeast (Wolfe and Shields 1997).

Identified in Fig. 1 are one of the better studied tetralogous regions in the human genome represented by 1q23, 6p21·E3, 9q34 and 19p13·E3. One provision here is that the human 1q23 region underwent tandem duplication and a subsequent inversion placed a copy of 1q23 on 1p13 (Katsanis et al. 1996; Kasahara et al. 1997). Another point to remember is that the genome project of either human or mouse is by no means near its completion. Accordingly, information currently available is extremely spotty. For example, of the four tetralogous regions identified in Fig. 1, 6p21·E3 region where the HLA major histocompatibility gene complex resides represents the most thoroughly analyzed of all the human chromosomal regions (Kasahara et al. 1996), whereas 19p13·E3 represents the exact opposite. It is for this reason that I have only listed in Table 2 one set of tetralogous genes and five sets of trilogous genes residing in these four tetralogous regions. It should be kept in mind that, excepting trilogous genes encoding calcium channels (CACNL1A6, ACCNL1A5 and CACNL1A4), fourth members of the other four sets of trilogues may still be found on 19p13·E3. In Table 3, however, aside from two further sets of trilogues, one unusual set of trilogues and two sets of dilogues are also shown. In addition, a set of genes encoding class II MHC antigens is shown as a series of monologues residing on the 6p21·E3 region only. Shown in the second row of Table 2 are trilogues residing in 1q23, 1p13 and 19p13·E3, encoding sodium and potassium ion ATPase alpha-chains. It should be recalled that 1p13 was a recent tandem duplicate of 1q23. Accordingly, the base sequence difference between ATP1A2 of 1q23 and ATP1A1 of 1p13 is considerably less than that between these two combined and ATP1A3 of 19p13·E3.

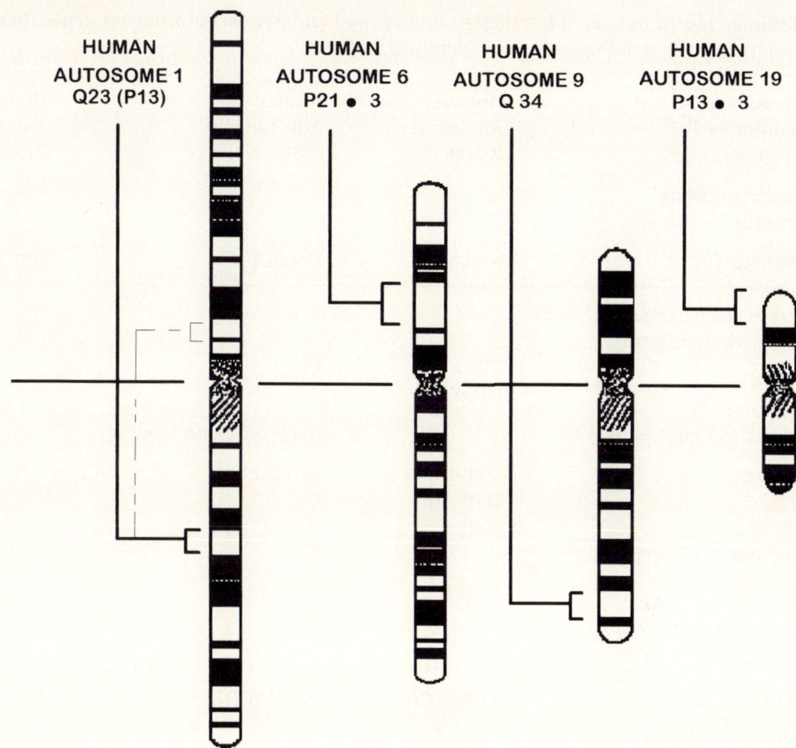

Fig. 1. One set of tetralogous regions in the human genome distributed over chromosomes No. 1, No. 6, No. 9 and No. 19 are physically identified on these four chromosomes schematically G-banded. As to the significance of the 1p13 region, indicated by *dotted lines*, see the text

The remaining pair of the tetralogous regions, namely 6p21·E3 and 9q34 regions, contain a pair of genes encoding ABC (ATP-binding cassette) transporters. Inasmuch as ATPases and ABC transporters share in common the specific binding affinity towards ATP, were it not for the known fact that ABC transporters were already present in invertebrate genomes, one might have made a mistaken assumption that a pair of ATPases and a pair of ABC transporters were encoded by original tetralogues.

8.3
Tetralogues, Trilogues and Dilogues that Still Contribute to Functional Redundancy

As the first tetraploidy event of the vertebrate genome occurred 540 million years ago, and the second event 330 million years ago, one might have thought that by now every one of the tetralogues, trilogues and dilogues has already

Table 2. List of one set of tetralogous (above line) and five sets of trilogous genes (below line) residing on the tetralogous regions identified in Fig. 1

Human Autosome 1 Q 23 (P13)	Human Autosome 6 P 21·3	Human Autosome 9 Q 34	Human Autosome 19 P13·3
Embryogenesis membrane proteins			
Notch 2 (P13)	Notch 4	Notch 1	Notch 3
Pre-B cell leukemia transcription factors			
PBX1	PBX2 (HOX12)	PBX3	
Heatshock protein HSP70		Ig Heavy chain binding	
HSP 6	HSP-HOM	GRP 74 6	
HSP 7	HSP AIL		
	HSP A		
Retionoid x Receptors			
RXR G	RXR B	RXR A	
Extracellular matrix proteins			
TENASCINR	TENASCINX	TENASCINC	
Calcium channel			
CACNL 1A6		CACNL1A5	CACNL1A4

differentiated to be endowed with its own unique function. Surprisingly, this has not necessarily been the case. It should be recalled that of the original tetralogues of Hox 3 gene loci, Hox A3, Hox B3 and Hox D3 still survive in vertebrates. Yet, seminal knockout experiments in the mouse revealed that neither a knockout of A3 nor of D3 resulted in a drastically altered phenotype; only when both Hox A3 and Hox D3 genes were simultaneously rendered nonfunctional, did the first cervical vertebra knwon as atlas disappear, in spite of the fact that such atlasless mutant mice still maintained the Hox B3 gene in the intact state (Condie and Cappechi 1994). It appears that Hox B3 gene has already forfeited its originally assigned role; if it is still performing a function it must be for an unkown subsidiary role. Whenever a new gene is found, it is now customary to perform a knockout experiment in the mouse to assess its functional significance. Sad experiences by many an investigator have been that such a knockout mouse more often than not manifested no specific defect. The situation described above is a testament to the long lasting effect of the two successive rounds of tetraploidy in the past that gave the characteristic redundancy to the vertebrate genome.

Table 3. Other sets of trilogous, dilogous and monologous genes residing within the same tetralogous regions. MR1 shown near the bottom left has been reported as a solitary class I MHC gene located in 1q25 region (Hashimoto et al. 1995). Inasmuch as 1q25 is outside of 1q23, one cannot be certain as to whether or not the dilogous relationship exists between this MR1 and a group of HLA class I genes residing within 6p21·E,R region

Human autosome 1 Q 23 (P13)	Human autosome 6 P 21·3	Human autosome 9 Q 34	Human autosome 19 P 13·3
Collagen alpha-chains COL11A1	COL11A2	COL5A1	
ATPase, Na$^+$, K$^+$ alpha-chains *ATP1A2* *ATP1A1* (P13)	ABC Transporter associated with antigen-processing *TAP1*	ABC (ATP-Binding cassette) transporter *ABC2*	ATPase, Na$^+$, K$^+$ alpha-chains *ATP1A3*
Complements	Class III MHC C2 C4A C4B	C5	C3
MR 1 (Q25)	Class I MHC HLA-A HLA-B HLA-C HLA-E HLA-F HLA-G		
	Class I MHC HAL-DR HLA-DQ HLA-DP etc.		

8.4
The Emergence of New Genes from Some of the Tetralogues Trilogues and Dilogues

Nevertheless, it is a fact that without knockout experiments, one can still point out several instances of new genes that emerged among the trilogues and dilogues listed in Tables 2 and 3. Shown in the third row of Table 2 are genes encoding HSP70 heat shock proteins. Those residing in 1q23 and 6p21.3 multiplied in numbers by local tandem duplications, whereas the one on 9q34 gained a specialized new function of specifically binding to immunoglobulin heavy chains, thus becoming the new gene GRP74.

We have already touched upon a pair of genes encoding ABC (ATP-binding cassette) transporters shown in the second row of Table 3. One residing in the 6p21·E3 region gained a specialized function to become a component of the antigen processing procedure, thus emerging as the new gene TAP1 (Kasahara et al. 1996). Again, the adaptive immune system is unique to gnathostomic (jawed) vertebrates.

Far more remarkable are trilogues encoding C4, C5 and C3 complements shown in the third row of Table 3, under the heading of CLASS III MHC. It has been known for some time that C3, C4 and C5 share 30 to 40% amino acid sequence identities among each other. Fortunately, this common ancestor of C3, C4 and C5 has recently been identified in a sea urchin belonging to the Cambrian phylum Echinodermata. This sea urchin protein, sharing 26 to 28% amino acid sequence identities with mammalian complements C3, C4 as well as C5, functioned simply as an opsonin (Smith et al. 1996). The opsonin tags invading pathogens to enhance phagocytic activity by host macrophages and granulocytes. As to the vertebrate complement system, the key event is a proteolytic conversion of the central component C3 into C3b, which covalently tags invading microorganisms to be cleared from the circulation or tissues. The C3 convertase of the classical pathway is composed of C2 and C4, whereas that of the alternative pathway is made of Bf and C3. In a relastionship similar to that enjoyed among C3, C4 and C5, C2 and Bf are also related to each other. Table 3 shows that in close linkage with C4, C2 also resides in the 6p21·E3 region. It follows that Bf might reasonably be localized within one of the three remaining tetralogous regions. However, recent information yet to be published indicates that, in teleost fish, a single protein still functions as both C2 and Bf. If so, differentiation of C2 from Bf is not due to the ancient tetralogous event, but due to a more recent tandem duplication. At any rate, there remains little doubt that the vertebrate complement system owes its existence to the tetralogous situations in the vertebrate genome created by the two successive rounds of tetraploidy events. The same thing can be said of genes encoding class I and class II HLA antigens located only in the 6p21·E3 region (Table 3). Peptide-presenting MHC antigens as instruments of self-nonself discrimination by the adaptive immune system are found only in gnathostomic (jawed) vertebrates. In order to be able to present self as well as nonself peptide fragments, the 90–110 amino acid-residue-long lg domain, originally made exclusively of antiparallel beta-strand structures, had to undergo a drastic transformation to gain one alpha-helical structure located atop these antiparallel beta-strands. Such a transformation succeeded only because there were tetralogues encoding successive lg-like domains to experiment with.

9

Conclusions

Life on this earth started 3.5 billion years ago in the CO_2 saturated Archaen ocean in the form of photosynthesizing cyanobacteria. Continuous release of molucular oxygen by these cyanobacteria finally created a reasonably aerobic environment about one billion years ago. At this point, our ancestral animal gained the ability to utilize molecular oxygen from an endosymbiotic Paracoccus-like bacterium that transformed itself to a mitochondrion. Inasmuch as animal mitochondria of diverse metazoan phyla have maintained an identical set of 37 genes, it is likely that as late as a few hundred million years before the Cambrian explosion, all the metazoan phyla of the future were still represented by one common ancestor that might yet have been unicellular. Inasmuch as diploblastic metazoa of the phyla Porifera and Cnidaria emerged immediately before the Cambrian explosion, during the brief Ediacaran episode, the Cambrian explosion sensu stricto involved triploblastic metazoa. Although Cambrian animals of 540 million years ago exclusively dwelt at the ocean bottom, practically all the extant animal phyla were represented. The nearly simultaneous emergence of all the metazoan phyla indicated that they shared the nearly identical genome. This genome is defined as the Cambrian pananimalia genome. Some of the more pertinent genes included in this pananimalia genome were discussed. The gnathostomic vertebrate genome became unique by undergoing two rounds of tetraploidy events at its inception. Thus, the tetralogous situation was created for each gene locus contained in the Cambrian pananimalia genome. Out of these tetralogues evolved new sets of genes that are to be found only in jawed vertebrates; e.g., genes for the adaptive immune system.

References

Attardi G (1988) Biogenesis of mitochondria. Annu Rev Cell Biol 4:289–333

Bui ETN, Bradley PJ, Johnson P (1996) A common evolutionary origin for mitochondria and hydrogenosomes. Proc Natl Acad Sci USA 93:9651–9656

Carroll SB (1995) Homeotic genes and the evolution of arthropods and chordates. Nature 376:479–485

Cavalier-Smith T (1987) The simultaneous symbiotic origin of mitochondria, chloroplasts and microbodies. Ann N York Acad Sci 503:55–71

Chang S (1994) The planetary setting of prebiotic evolution. In: Bengston S (ed) Early life on earth. Nobel symposium no 84. Columbia University Press, New York

Chen J-Y, Dzik J, Edgecomb GD, Ramskold L, Zhou G-Q (1995) A possible early Cambrian chordate. Nature 377:720–722

Cloud P, Glaesner MF (1982) The Ediacaran period and system: Metazoa inherit the earth. Science 218:783–792

Collura RV, Stewart C-V (1995) Insertions and duplications of mtDNA in the nuclear genomes of old world monkeys and hominoids. Nature 378:485–489

Condie BG, Capecchi MR (1994) Mice with targeted disruptions in the paralogous genes HOXa3 and HOXd-3 revealed synergistic interactions. Nature 370:304–307

Conway Morris S (1989) Burgess shale faunas and the Cambrian explosion. Science 246:339–346

Davis AP, Witte DP, Hsieh-Li HM, Patter SS, Capecchi M (1995) Absence of radius and ulna in mice lacking hoxa-11 and hoxd-11. Nature 375:791–795

De Duve C (1991) Blueprint for a cell: the nature and origin of life. Carolina Biological Supply, Burlington

Degens ET, Kempe S, Spitzy A (1984) CO$_2$: a biogeochemical portrait. In: Huntzinger O (ed) Handbook of environmental chemistry. Springer, Berlin Heidelberg New York

Escriva H, Safi R, Hanni C, Langlois M-C, Saumitou-Laprade P, Stehelin D, Capron A, Pierce R, Laudet V (1997) Ligand binding was acquired during evolution of nuclear receptors. Proc Natl Acad Sci USA 94:6803–6808

Gabbot SE, Aldridge RJ, Theron JM (1995) A giant conodont with preserved muscles from the upper Ordovician of South Africa. Nature 374:800–803

Gould SJ (1995) Of it and not above it. Nature 377:681–682

Hashimoto K, Hirai M, Kurosawa Y (1995) A gene outside the human MHC related to classical HLA class I genes. Science 269:693–695

Holland PWH, Garcia-Fernandez J, Williams NA, Sidow A (1994) Gene duplication and the origins of vertebrate development. Development Suppl [no volume number]:125–133

Jacobs HT, Elliott DJ, Math VB, Farquharson A (1988) Nucleotide sequence and gene organization of sea urchin mitochondrial DNA. J Mol Biol 202:185–217

Kasahara M, Hayashi M, Tanaka K, Inoko H, Sugaya K, Ikemura T, Ishibashi T (1996) Chromosomal localization of the proteasome Z subunit gene reveals an ancient chromosomal duplication involving the major histocompatibility complex. Proc Natl Acad Sci USA 93:9096–9101

Kasahara M, Nakata J, Sattay, Takahata N (1997) Chromosomal duplication and the emergence of the adaptive immune system. Trends Genet 13:90–92

Katsanis N, Fitzgibbon J, Fisher EMC (1996) Paralogy mapping: identification of a region in the human MHC triplicated onto human chromosomes 1 and 9 allows the prediction and isolation of novel PBX and NOTCH loci. Genomics 35:101–108

Kempe S, Degens ET (1985) An early soda water ocean? Chem Geol 53:95–104

Lavorga G, Ueda H, Clos J, Wu C (1991) FTZ-F1, a steroid hormone receptor-like protein implicated in the activation of fushitarazu. Science 252:848–851

Lewis EB (1978) A gene complex controlling segmentation in Drosophila. Nature 276:567–570

Loosli F, Kmita-Cunisse M, Gehring W (1996) Isolation of a Pax-6 homolog from the ribbon worm *Lineus sanguineus* expressed in eyes. Proc Natl Acad Sci USA 93:2658–2663

Lundin LG (1993) Evolution of the vertebrate genome as reflected in paralogous chromosomal regions in man and the house mouse. Genomics 16:1–19

Luo X, Ikeda Y, Parker KL (1994) A cell specific nuclear receptor is essential for adrenal and gonadal development and sexual differentiation. Cell 77:481–490

Matsuo T, Osumi-Yamashita N, Noji S, Ohuchi H, Koyama E, Myokai F, Matsuo N, Taniguchi S, Doi H, Iseki S, Ninomiya, Y, Fujiwara M, Watanabe T, Eto K (1993) A mutation in the Pax-6 gene in rat small eye is associated with impaired migration of midbrain crest cells. Nature [Genet] 3:299–304

McGinnis N, Kuziora MA, McGinnis W (1990) Human Hox4-2 and *Drosophila* deformed encode similar regulatory specificities in *Drosophila* embryos and larvae. Cell 63:969–979

Ohno S (1970) Evolution by gene duplication. Springer, Berlin Heidelberg New York

Ohno S (1972) So much junk DNA in our genome. In: Smith HH (ed) Evolution of genetic systems. Brookhaven symposium no 26. Gordon and Breach, New York

Ohno S (1996) The notion of the Cambrian pananimalia genome. Proc Natl Acad Sci USA 93:8475–8478

Schopf JW (1993) Microfossils of the early archean apex chert: new evidence of the antiquity of life. Science 260:640–646

Smith LC, Chang L, Britten RJ, Davidson EH (1996) Sea urchin genes expressed in activated coelomocytes identified by expressed sequence tags. J Immunol 156:593–602

Spring J (1997) Vertebrate evolution by interspecific hybridization – are we polyploid? FEBS Lett 400:2–8

Sugaya K, Fukagawa T, Matsumoto K-I, Mita K, Takahashi E-I, Ando A, Inoko H, Ikemura T (1994) Three genes in the human MHC class III region near junction with the class II: gene for recetor of advanced glycosylation end products, PBX2 homeobox gene and a Notch homolog, human counterpart of mouse mammary tumor gene int-3. Genomics 23:408–419

Sun H, Rodin A, Zhou Y, Dickinson H, Harpe DE, Hewett Emmett D, Li WH (1997) Evolution of paired domains: isolation and sequencing of jellyfish and hydra Pax genes related to Pax5 and Pax6. Proc Natl Acad Sci USA 94:5156–5161

Urbanek P, Fetka I, Meisler H, Busslinger M (1997) Cooperation of Pax2, Pax5 in midbrain and cerebellum development. Proc Natl Acad Sci USA 94:5703–5708

Van Verseveld HW, Bosma G (1987) The respiratory chain and energy conservation in the mitochondrion-like bacterium *Paracoccus denitrificans*. Microbiol Sci 4:329–333

Wolfe KH, Shields DC (1997) Molecular evidence for an ancient duplication of the entire yeast genome. Nature 387:708–713

Wolstenholme DR, MacFarlane JL, Okimoto R, Clary DO, Wahleithner JA (1987) Bizzare tRNAs inferred from DNA sequences of mitochondrial genome of nematode worms. Proc Natl Acad Sci USA 84:1324–1328

Zischler H, Geisert H, von Haeseler A, Paabo S (1995) A nuclear "fossil" of the mitochondrial D-loop and the origin of modern humans. Nature 378:489–492

Evolution of Metazoan Collagens

R. Garrone[1]

1
Introduction: An Up-To-Date Definition of Collagen

1.1
The Collagen Molecule

The word "collagen" was used in the 19th century to characterize a constituent of connective tissue (mainly skin or bones) which yields gelatin on boiling. The staining properties of collagen made it obviously visible on tissue sections. From this histological detection in human and vertebrate organs, it was suggested that at least some invertebrates actually contain the same kind of substance. The development of electron microscopy and the improvement of biochemical analyses led to the clear localization of collagen in both vertebrates and invertebrates (Piez and Gross 1959). However, most studies focused on the easily recognizable collagen fibrils. Later, it became evident that there was a family of collagenous proteins and that the definition had to be clarified (for a review see van der Rest and Garrone 1990, 1991; van der Rest et al. 1993).

The current definition takes into account first the triple helical domain of the collagen molecule. Each collagen molecule is formed by three subunits, the α chains, which share unique properties. They contain amino acid sequences repeating the triplet Gly-Xaa-Yaa, where Xaa and/or Yaa are frequently the imino acid proline. This makes a characteristic helical conformation of the peptide, the collagen α helix, with a tendancy to assemble with similar peptides to constitute a coiled-coil triple helix (Brodsky and Ramshaw 1997), the collagenous domain. The three α chains constituting the molecule can be the same (three $\alpha1$), or not (two $\alpha1$ and one $\alpha2$, or $\alpha1$, $\alpha2$, $\alpha3$). The resulting structure has a high stability, due to the position of glycine (a small amino acid) close to the axis of the helix, and due to the stabilizing action of proline via different kinds of hydrogen bonds (Bella et al. 1994; Bella and Berman 1996). In addition, this type of molecular organization includes water molecules (Bella et al. 1995). The collagen triple helix is resistant to temperature denaturation (the melting temperature of collagen is above the body or environmental tempera-

[1]CNRS Institute of Biology and Chemistry of Proteins, Claude Bernard University, Lyons, France

Progress in Molecular and Subcellular Biology, Vol. 21
W.E.G. Müller (Ed.)
© Springer-Verlag Berlin Heidelberg 1998

ture) and to enzymatic cleavage (except for the specific collagenases). During the process of the collagen α helix synthesis, some the the prolyl residues are hydroxylated into hydroprolyl residues, a modification which enhances the stability of the molecule and which seems necessary for a correct assembly of the three α chains (Mazzorana et al. 1993). In the animal kingdom, hydroxyproline is then highly specific of the protein collagen. Similarly, some of the lysyl residues are hydroxylated into hydroxylysyl residues which can be further glycosylated by addition of a monosaccharide (galactose), eventually completed with glucose (galactosyl-glucose). In vertebrates at least, the two hydroxylases are noticeable. The prolyl 4-hydroxylase is a tetrameric protein containing two pairs of subunits $\alpha_2 \beta_2$. Curiously, the β subunit is a multifunctional polypeptide: it shares identity with the protein (disulfide isomerase) which catalyzes the formation of disulfide bonds (Koivu et al. 1987) and can have several other functions (Kivirikko et al. 1989). Interestingly, the disulfide isomerase forms high-order complexes with newly synthesized collagen in the lumen of the endoplasmic reticulum (Kellokumpu et al. 1997). The lysyl hydroxylase is a resident rough endoplasmic luminal protein, associated with the membrane but lacking any known endoplasmic reticulum-specific retention motifs (Kellokumpu et al. 1994).

The triple helical (or collagenous) domain is flanked and interrupted by non-triple helical (or non-collagenous) domains. This structure reflects the fact that the sequence of each chain within these regions has no repetition of the triplet Gly-Xaa-Yaa. Actually, the corresponding domains often have a modular organization (Bork 1992; Brown and Timpl 1995): they are made of one or more sequences found in other proteins. In fact, this is a characteristic of many proteins of the extracellular matrix (Engel 1991) and even of many other proteins (Bork et al. 1996). Finally, the triple-helix motif also is encountered in non-collagenous proteins (Brodsky and Shah 1995). This latter fact completes the definition of collagen. In addition to containing at least one triple helical domain (defined as above), a collagenous protein must be involved in the organization of the extracellular matrix. This definition excludes several proteins which contain a collagenous domain, but fulfil other functions: the C1q component of the complement system, lung surfactant proteins, mannose binding proteins and some circulating lectins (collectins), some forms of acetylcholinesterase, the macrophage scavenger receptor etc. (Prockop and Kivirikko 1995).

1.2
The Collagen Family

In vertebrates, such a definition has allowed the characterization of a large family of proteins: 19 types of collagen have been described which are coded by at least 33 different genes, and responsible for a wide spectrum of diseases (Prockop and Kivirikko 1995). Several subtypes have been established and the major divisions occur between the collagens which are able to form banded fibrils (the banding pattern having 67 nm in periodicity), the fibrillar collagen types I, II, III, V and XI (which will be detailed in the Sect. 2) and all the others, the so-called non-fibrillar collagens. Among the latter, there are the collagens involved in the framework of basement membranes, the type IV (to be reviewed in Sect. 3), and other types, some of them being known almost only by their cDNA (i.e. types XV and XVIII, which are thought to be close to the basement membrane area).

Some of them will be studied in Sections 4 and 5 of this chapter, nevertheless it might be important to emphasize the following peculiar points:

1. A large group comprises collagen molecules which do not assemble together (as all the others do) but the molecules of which are localized at the surface of fibrillar collagen assemblies. These are the FACIT collagens (for Fibril Associated Collagens with Interrupted Triple helices) including types IX, XII, XIV, and probably types XVI and XIX (Prockop and Kivirikko 1995).

2. The type VI collagen forms beaded filaments, observed in transmission electron microscopy as microfibrils with a transverse striation of 100 nm periodicity and encountered in almost all tissues (Bruns 1984). It is remarkable because it contains a very short (105 nm) triple helical domain for a collagen molecule, and one of its chains, the $\alpha 3$, is about twice the mass of each two other type VI α chains. In addition, the C-terminus of this chain possesses a Kunitz-type domain (similar to a sequence found in serine protease inhibitors of the Kunitz type), the three-dimensional structure of which has been clarified only recently by X-ray diffraction after crystallization (Arnoux et al. 1995) and a nuclear magnetic resonance study in solution (Zweckstetter et al. 1996). This emphasizes the fact that even for vertebrate collagens, the structure of some non-triple helical parts is still poorly understood.

3. Type VII collagen has a restricted tissue distribution, being associated with the basement membrane zone of stratified squamous epithelia. It is the longest vertebrate collagen molecule (424 nm compared with the 300 nm of fibrillar collagen molecules). It forms antiparallel dimers, with an overlap of 60 nm, which assemble laterally to built anchoring fibrils (Sakai et al. 1986). The association of type VII collagen with an isoform of laminin, laminin 5, can form anchoring complexes (Leivo et al. 1996). The C-terminal domain of the $\alpha 1$(VII) chain (the molecule is an homotrimer) contains a Kunitz-type motif, as for the $\alpha 3$(VI) chain (Greenspan 1993). A noticeable characteristic of type VII collagen is the organization of its gene. It contains 118 exons, the highest number of any previously characterized gene (Christiano et al. 1994).

4. The collagen types VIII and X are similar short-chain collagens. Their molecules (a heterotrimer for type VIII, a homotrimer for type X) can self-assemble via their C-termini to form highly regular, hexagonal networks (Sawada et al. 1990; Kwan et al. 1991). These arrays are encountered in

basement membranes for type VIII (the Descemet's membrane at the posterior face of the cornea is the most striking example). Type VIII collagen is synthesized by endothelial and also smooth muscle cells (MacBeath et al. 1996). Type X collagen is encountered in the territorial matrix of hypertrophic chondrocytes (Lu Valle et al. 1993), and it can also move through the cartilage matrix and associate with pre-existing type II collagen fibrils (Chen et al. 1990). Type X collagen has a role during endochondral bone formation (Wallis 1993). These two collagens share a condensed genomic organization completely distinct from all other vertebrate collagen genes, with only three exons. In particular, only one large exon encodes a part of the N-terminal, non-helical domain, and the entire helical and non-helical C-terminal domains (Muragaki et al. 1991; Kong et al. 1993).

5. Finally, two new collagens (types XIII and XVII) deserve a special mention as they contain a transmembrane domain. Type XIII collagen belongs to the group of non-fibrillar, short-chain collagens and several hundred forms are known due to complex alternative splicing of the primary transcript (Pihlajaniemi and Tamminen 1990). There is now some evidence that at least some forms of type XIII collagen molecules have a transmembrane region in the non-collagenous N-terminal domain. This collagen, which interacts intracellularly with talin, vinculin and the β3 integrin subunit could be a focal adhesion protein with an adhesive function (Väisänen et al. 1996). A similar role has been suggested for type XVII collagen, which has a transmembrane domain (Li et al. 1993). It interacts with the α6 integrin subunit (Hopkinson et al. 1995) and is present in the hemidesmosomes of epidermal tissues (Li et al. 1991).

2
Collagen Fibrils: From Sponges to Humans

2.1
The Homogeneous Subgroup of Fibrilar Collagen in Vertebrates

The molecules belonging to the subfamily of fibrillar collagen have three characteristics: they have a long (about 300 nm), uninterrupted triple helical domain; a highly conserved non-collagenous C-terminal domain, and they can assemble laterally, with a constant shift (about a quarter of their length; van der Rest and Garrone 1990, 1991; van der Rest et al. 1993). This association builds long cylinders and the regularity of the polymerization creates the transverse alignment of similarly charged residues. Hence, with the processing of samples for electron microscopy (positive staining) a transverse striated pattern becomes visible; it has a motif repeated every 67 nm: the period of the collagen fibril. Actually, with other processing methods (negative staining), alternating clear and dark bands are visible; they correspond to dense and weak molecular packing regions respectively, but the motif (a clear band plus a dark band) has again a length of 67 nm. Atomic force microscope studies have shown that this periodic pattern is not an electron microscope artefact (Revenko et al. 1994). The molecules of the fibrillar collagens are processed before their assembly. When the molecules are secreted (eventually after an intracellular pre-assembly), the non-collagenous C-terminal domain is rapidly cleaved by a specific endoprotease. The enzyme which cleaves the C-terminal domain of collagen types I, II and III is identical to the bone morphogenetic protein 1 (BMP 1), a metalloprotease with a bone-inductive activity. This enzyme is also involved in the maturation of the lysyl oxidase (Cronshaw et al. 1995) which cross-links collagen molecules. The fact that other BMPs are members of the transforming growth factor-β (TGF-β) superfamily raises the possibility that BMP 1 represents a crossroad between the cellular response to cytokines and the assembly of the extracellular matrix (Olsen 1996). The N-terminal non-collagenous domain is more variable among fibrillar collagens. Even its processing can be different: rapidly cleaved in type I and II collagens, it is more slowly released in type III and can remain, at least partially, in type V and XI collagens.

Although collagen fibrils can be reconstituted in vitro from each of the fibrillar collagen types, the important point is that the fibrils encountered in tissues are biological alloys made of several collagens. In addition, the ratio of the so-called minor collagens, the highly similar types V and XI, can control the fibril diameter, most likely via their N-terminal extensions which remain: mutations in two genes, *COL5A2* and *COL11A1* coding for α chains of these collagens result in increased collagen fibril diameters (Andrikopoulos et al. 1995; Li et al. 1995).

2.2
The Primitive Fibrillar Collagens

Collagen fibrils which are very similar to vertebrate collagen fibrils have been observed in several invertebrates. In fact, three possibilities occur: (1) fibrils which have a complex banding pattern, more or less similar to that of vertebrates, with a variable diameter; (2) fibrils which have a simplified banding pattern and usually a thin diameter (about 20 to 25 nm); and (3) no structure which can be identified as a collagen fibril, at least with morphological methods. This last case is not frequent, examples being rotifers, *Drosophila* (Garrone 1978) and most likely the nematodes (Cox 1992; Kramer 1994).

In invertebrates, there is obviously a gap between morphological data and molecular evidence. Although very few species have been investigated at the molecular level, as the data are consistent, a tentative scheme of evolution can be drawn.

In sponges (Gross et al. 1956; Garrone et al. 1975; Garrone 1978, 1984, 1985; Diehl-Seifert et al. 1985; Gramzow et al. 1988) and in cnidarians (see Franc 1985 for a review; Weber and Schmid 1985; Tillet et al. 1996), thin fibrils (20 to 25 nm in diameter) are described. They show a simplified banding pattern (a period estimated around 60 nm, with usually three subperiods), clearly visible often only in negatively stained or rotary shadowed preparations from isolated fibrils. A remarkable exception has been reported by Ledger (1974) in calcareous sponges where giant fibres (up to 130 nm in diameter) are seen. However, these structures might result from the aggregation of thinner filaments belonging to another collagen type present in sponges (the spongin or short-chain collagen). Small microfibrils are often encountered, specially in cnidarians. Their nature is controversial and it is not sure at all that they always represent collagenous structures. On the contrary, at least in one clear situation, the mesogleal microfibrils of a jellyfish are formed by fibrillin, another protein of the extracellular matrix (Reber-Müller et al. 1995). These primitive collagen fibrils have been investigated in one of two ways, depending on the type of material.

In sponges, where the collagen is largely insoluble (Garrone 1978), a fruitful approach has been molecular biology. The results obtained at the cDNA level led to the partial characterization of a putative collagenous chain constituted by 307 amino acid residues of a helical domain, with one imperfection, and a C-terminal non-helical domain containing 240 amino acid residues including 7 cysteines. Amino acid sequence comparisons indicated that this sponge collagen is homologous to vertebrate fibrillar collagen (Exposito and Garrone 1990). The high conservation of this non-helical C-terminal domain (with the cysteines similarly located) in all the fibrillar collagen chains facilitates this identification. In vertebrates, it is assumed that the chain assembly, which takes place in the cell endoplasmic reticulum, is under the control of the C-terminal domain, a hypothesis which fits well with the constancy of this domain.

Additional investigations at the genomic level confirmed that this chain belonged to a fibrillar collagen (with information on the N-terminal part of the chain) and showed two remarkable features of gene organization. Firstly, the general organization of the gene coding for the fibrillar sponge collagen is very similar to the organization of the vertebrate fibrillar collagen genes (Exposito et al. 1993b). The main triple helical domain is encoded by 31 exons, most of them being related to a 54 base-pair (bp) unit and beginning with an intact glycine codon. Considering all the exon/intron junctions of the fibrillar collagen genes, an ancestral collagen gene has been reconstituted. It is suggested that it arose from multiple duplications of a 54-bp exon and a (54 + 45)-bp module (Exposito et al. 1993) A similar hypothesis has been put forward again recently (Takahara et al. 1995; Vuoristo et al. 1995). The second interesting point is the possible relationships between sponge fibrillar collagen and vertebrate minor fibrillar collagens (types V and XI). The comparison of sequences with type XI collagen was highly suggestive, and an additional similarity was observed at the end of the triple helical coding region for these two collagens. Within a junction exon of type XI collagen, there are 18 bp coding for the end of the helical domain. There is no junction exon in the sponge collagen gene, but a single 18 bp exon coding for the end of the helical domain (Exposito and Garrone 1990).

In cnidarians the collagen is soluble in acetic acid after pepsin treatment and a biochemical approach has been used. The more recent data emphasized, here again, the relationships between these primitive collagens and the minor vertebrate collagen type V (Tillet-Barret et al. 1992; Tillet et al. 1996). Electron microscope observation of rotary-shadowed collagen fibrils from jellyfish reveals protrusions extending perpendicularly to the fibrils (Reber-Müller, pers. comm.), which might correspond to the remnants of N-terminal non-helical domains of the collagen molecule, a delayed cleavage again comparable to the situation for minor fibrillar collagens.

These observations lead to the hypothesis that vertebrate type V and XI fibrillar collagens are the most ancient collagens. They are only able to form thin fibrils (like primitive collagen fibrils) and they might constitute some kind of axial template, over which the major collagens could polymerize (types I and III mixed with type V; type II mixed with type XI).

Simplified collagen fibrils can be encountered in zoological groups from sponges to nemertines (Pedersen 1968). But collagen fibrils with a more complex banding pattern are only visible in more evolved zoological groups. Primitive-type fibrils can be mixed with more complex fibrils. This seems the case in sea urchin embryos, where thin fibrils (25 nm) are observed in association with the larval skeleton (Lethias et al. 1997). Mechanical isolation of these fibrils and electron microscope observation after rotary-shadowing reveals that they have a "lampbrush" aspect, as they are decorated with many filaments pointing out perpendicularly to the fibre surface. Characterization of the corresponding cDNA (Exposito et al. 1992a) and gene (Exposito et al. 1995) reveals that at least one chain of the collagen molecules constituting such fibrils has the characteristic C-terminal and long helical domain of a fibrillar collagen, but its N-terminal domain is unique (Exposito et al. 1992) and contains 12 repeats of a novel protein module (Exposito et al. 1995). Antibodies have been raised against recombinant peptides of this peculiar N-terminal domain. They showed that the N-propeptide of this collagen is retained in mature fibrils and corresponds to the extensions visible at the surface of isolated fibrils (Lethias et al. 1997).

2.3
The Vertebrate-Type Fibrillar Collagens

In molluscs, where perfectly organized fibrils are seen with electron microscopy, homotrimeric (Schmut et al. 1980) and heterotrimeric (Kimura et al. 1981) molecules have been described, and this discrepancy is probably due to the subgroup studied. An interesting collagen fibril differentiation is found in the squid. While the connective tissue fibrils have a wide diameter and show 12–13 intraperiodic electron-dense bands (Bairati et al. 1995), the head "cartilage" contains thin collagen fibrils (Bairati et al. 1987). However, when these fibrils are isolated from the tissues, they show a periodic structure similar to that of vertebrate type I collagen fibrils (Bairati et al. 1989).

Morphological and biochemical studies were done on some annelid interstitial collagen (Hermans 1970; Vitellaro-Zuccarello et al. 1985), and revealed the complex banding pattern of the fibrils. However, detailed sequence data were missing. Partial cDNA sequences corresponding to fibrillar collagen chains have been cloned recently in the annelids *Arenicola marina* (Sicot et al. 1997) and *Alvinella pompejana* (Sicot, unpubl. data). The C-terminal non-collagenous domain appears highly homologous to the other known similar domains in fibrillar collagen chains. However, it contains only six cysteines at conserved positions (compared with the usual number of seven or eight), the cysteines C2 and C3 are missing. This peculiarity might mean that the cysteines 2 and 3 of all the fibrillar collagen chains are not involved in inter-chain cross-links within the molecule and emphasizes the interest in studying invertebrate collagens. Within the collagenous domain, a frequent Gly-Gly repeat is noted. This is due to the fact that the amino acid glycine is often in the Xaa or Yaa position in the triplet Gly-Xaa-Yaa, a seemingly characteristic sequence of invertebrate fibrillar collagens, although this is not the case for the fibrillar collagen of *A. pompejana* (Sicot, unpubl. data). A recognition site of lysil oxidase is observed, an uncommon feature in invertebrate collagens (Sicot et al. 1997). The data obtained by molecular biology are in good agreement with the biochemical analyses of fibrillar collagen obtained from the annelids *A. pompejana, A. caudata, Paralvinella grassei* and *Nereis diversicolor* (Gaill et al. 1995) as well as from the tube worm *Riftia pachyptila* (Gaill et al. 1991; Mann et al. 1992). The determination of temperature denaturation of these collagens points to a striking difference between the collagens of shallow seawater annelids (about 28 °C) or the species living close to deep-sea hydrothermal vents (from 35 to 45 °C; Gaill et al. 1995). Sequence studies from the fibrillar collagen of *A. pompejana* points to the high occurrence of the imino acid proline in Yaa position (a favourable position for further hydroxylation) and the high number of Gly-Pro-Pro triplets (Sicot, unpubl. data). Considering the stabilizing function of proline and hydroxyproline, this kind of sequence characteristic could explain the thermal stability of these collagens.

Few new data have been obtained concerning fibrillar collagen in arthropods in general and in insects in particular (for a review see François 1985). Two genetically distinct types of collagen have been demonstrated by biochemistry in crustaceans (Mizuta et al. 1994). A collagen-like sequence has been detected in insects, e.g. *Bombyx mori* (Chareyre et al. 1996).

Much more data are available for echinoderm collagen fibrils. Several studies have been done in the past (for a review see Bailey 1985) and they have been added to by more recent data. Collagen fibrils with a chain composition identical to vertebrate type I collagen have been characterized by biochemistry in the test (Omura et al. 1996) and spine ligament (Trotter and Koob 1994) of several sea urchins and starfish (Kimura et al. 1993). Two fibrillar collagen chains have been demonstrated by cDNA cloning in the body of the sea urchins *Paracentrotus lividus* and *Strongylocentrotus purpuratus* (D'Alessio et al. 1990; Exposito et al. 1992b). These data emphasize again the similarity between fibrillar sea urchin collagen and vertebrate type I collagen. However, the echinoderm collagen has special organization and function. The fibrils are often massively interacting with proteoglycans, some of them, at least, are strictly symetrical and they are contained in a tissue which has mutable properties, i.e. it can undergo rapid and dramatic alterations in its mechanical properties. This tissue is actually under physiological control by the nervous system. In a very short time scale (within seconds) it can stiffen or soften. The interactions with proteoglycans has been extensively demonstrated, especially in the sea cucumber (Trotter et al. 1995) but also in sea urchins (Trotter and Koob 1989) and sea lilies (crinoids; Erlinger et al. 1993). The collagen fibrils of echinoderms have been clearly demonstrated to be curvinearly tapered, more irregular in cross section in sea cucumbers than in sea urchins (Trotter et al. 1994), and frequently molecularly strictly bipolar (Thurmond and Trotter 1994). This shape has been related to the mutable properties of this collagenous tissue.

An interesting result has been obtained on chaetognath connective tissue (Duvert and Salat 1990). Species of this curious *incertae sedis*, planktonic group contains collagen fibrils with a variable diameter (up to 45nm), an irregular cross-section and a complex-type banding pattern (Duvert and Salat 1990).

3
Basement Membrane Collagen:
The Marker of Tissue Differentiation

3.1
The Basement Membrane Collagens of Vertebrates

The collagen type IV forms the scaffold of basement membranes (for a review see Kühn 1994; Timpl 1996). Its molecule has three main features: (1) the long (400nm) triple helical domain contains many short interruptions and then is very flexible; (2) its N-terminal end (the so-called 7S domain) can interact laterally in an antiparallel organization with other type IV collagen molecules, making spider-like tetramers; and (3) the C-terminal end, highly conserved among species, forms a knob interacting with similar domains of other tetramers. The result is a network comprising several layers and containing other molecules: laminins, nidogen, low molecular weight heparan sulfate proteoglycans, and several other proteins (BM 40, SPARC and others).

In vertebrates, six α chains have been characterized to form a type IV collagen molecule. They are divided into two homologous groups, the even (α1, α3, α5) and the uneven (α2, α4, α6) chains. The likely stoichiometries of the molecules are [α1(IV)]2α2(IV); [α3(IV)]2α4(IV) and [α5(IV)]2α6(IV), but other assemblies cannot be excluded. The corresponding human genes are found in pairs (*COL4A1-COL4A2* on chromosome 13; *COL4A3-COL4A4* on chromosome 2; *COL4A5-COL4A6* on chromosome X) and have head-to-head orientations with overlapping promoter regions (Hudson et al. 1993).

3.2
The Basement Membrane Collagens of Invertebrates

The vertebrate chromosomic arrangement of type IV collagen genes does not seem to be general. In nematodes, two type IV collagen chains have been detected (for a review, see Cox 1992; Kramer 1994), they are homologous to the $\alpha1$(IV) and $\alpha2$(IV) chains and their genes are located on different chromosomes (Guo and Kramer 1989). The $\alpha1$(IV) chain is encoded by the *clb-2* gene (locus *emb-9* located on chromosome III) and the $\alpha2$(IV) chain is encoded by the *clb-1* gene (locus *let-2* on the X chromosome; Sibley et al. 1994). Transcripts of the *C. elegans* $\alpha2$(IV) gene are alternatively spliced and the expression of the variants is developmentally regulated (Sibley et al. 1993). This is also the case for the *Ascaris suum* $\alpha2$(IV) gene transcript (Pettitt and Kingston 1994), a nematode collagen which has been extensively studied by biochemistry (Noelken et al. 1986). As only two α(IV) chains are likely to be present in *C. elegans*, this differential expression would suggest that the embryos may require a different form of collagen basement membrane than do adults (Sibley et al. 1993). Mutations affecting this basement membrane collagen lead to alterations of the phenotype (Sibley et al. 1994) and even to embryonic lethality (Guo et al. 1991). An $\alpha2$(IV) collagen gene from the filarial nematode *Brugia malayi* has also been characterized (Caulagi and Rajan 1995).

Two chains, designated 3α (Exposito et al. 1993a) and 4α (Exposito et al. 1994) of basement membrane collagens have also been characterized in sea urchins. They are respectively homologous to the $\alpha1$ and $\alpha2$ type IV vertebrate chains. The presence of at least two different α(IV) chains in invertebrates is now strongly established, as a second chain has been recently detected in *Drosophila* (Yasothornsrikul et al. 1997). The first *Drosophila* type IV collagen chain (see Fessler and Fessler 1989, for a review) has been known since 1987 (Blumberg et al. 1987; Cecchini et al. 1987). The lateral interactions of the molecules might be more extended than in vertebrates and the molecule is cleaved during metamorphosis (Fessler et al. 1993).

A basement membrane-like structure is almost always visible in cnidarians, separating epithelial bilayers from the mesoglea (the intervening, often acellular layer). A biochemical characterization of type IV collagen has been successful in *Hydra vulgaris* (Sarras et al. 1991).

3.3
The Origin of Basement Membrane Collagens

Basement membranes are not detectable on electron microscope micrographs of tissues from most sponge species. The usual organization shows the epithelia (termed pinacoderm for the whole cells surrounding the animal or lining the canals, or choanoderm for the internal flagellated cell assemblies) in direct contact with collagen fibrils. However, in only one group of marine sponges, the Homosclerophorida, a feltwork layer underlines all the epithelia. Interestingly, these sponges differ from all other species by producing a spermatozoon with an acrosome and they do have a complex extracellular matrix containing the adhesive protein tenascin (Humbert-David and Garrone 1993). From a species, *Pseudocorticium jarrei*, belonging to the Homoscleromorpha, cDNA and genomic studies detected two type IV collagen chains (Boute et al. 1996). One of them has been extensively characterized and appears highly conserved. The twelve cysteines are perfectly located in the C-terminal domain of the chain; most imperfections of the triple helical domain are in conserved positions and even the partly characterized exon/intron organization of the gene has the same features as for the other type IV collagen genes (Boute et al. 1996).

It can be assumed then that type IV collagen is the only ubiquitous collagen and that at least two genetically different chains are always present. The fact that most sponges are probably missing type IV collagen should be related to the poorly differentiated state of sponge tissues. It is well known that dissociation of sponge cells leads to a loss of cell specializations, and further reconstitution of a new organism by cell aggregation leads to specialization. The well-organized epithelium of Homosclerophorida, as an "evolved" character, probably represents the first stage of tissue differentiation during evolution. Basement membranes might strengthen this differentiation by a stabilizing effect via interactions with cells.

4
The Species-Specific Collagens

In addition to the conserved collagen fibrils and basement membrane collagens, peculiar collagens exist in some groups or even species. As they generally vary among species, even when they characterize a whole group or a phylum, the term species-specific will be used. A general review can be found in Bairati and Garrone (1985), with only some new developments detailed here.

4.1
The Mini-Collagens in Cnidarians

Most of these collagens, if not all, are secreted and located outside of the animal body. The only noticeable exception is found in the capsule of nematocysts. These defensive, explosive organelles are characteristic of cnidarians. They are contained within cells, the nematocytes, and they consist of a stiff wall, the capsule, enclosing a long, armed tube which can be everted following cell stimulation. Biochemical studies have demonstrated that the capsule of nematocysts contained cysteine-rich, low molecular weight collagenous proteins (Brand et al. 1993). Four collagen-related cDNA clones are highly expressed in developing nematocytes in hydra (Kurz et al. 1991). The corresponding deduced amino acid sequences revealed a central domain containing 14 to 16 Gly-Xaa-Yaa triplets, flanked by stretches of several proline residues. From these sequences a model has been proposed consisting of a 12–14nm long collagen triple helix bordered by polyproline II type helices (Kurz et al. 1991). Atomic force microscopy and field emission scanning electron microscopy studies of the capsule have demonstrated the presence of periodic fibre arrangements within the inner part of the capsule wall (Holstein et al. 1994). It has been suggested that this organization of a special collagenous structure can provide the tensile strength which is necessary to withstand the osmotic pressure in the capsule (Holstein et al. 1994). The capsule of nematocysts might have been a unique example of a mature intracellular collagenous protein, but since the discovery of plasma membrane-bound collagens (types XIII and XVII) this no longer seems to be exceptional.

4.2
The "Externally-Secreted" Collagens

The other species-specific collagens, which are all "external" collagens, can be divided into three groups: (1) the collagens secreted by a glandular tissue; (2) the collagens forming coarse fibers arranged in a plywood manner, and (3) the microfibrillar, short-chain collagens. The first group is represented in lower vertebrates by the egg case of the dogfish (for a review see Hunt 1985; Knight et al. 1996) and in invertebrates by the byssus filaments of the molluscs Bivalvia (see Bairati 1985). As no recent data are available, they will not be considered in this review.

4.2.1
Annelid Cuticles

The second category refers mainly to the cuticle of annelids and related species (from vestimentiferan tube worms). They are formed by the superimposition of successive layers of twisted coarse fibers with no visible banding pattern. The special organization (Gaill and Bouligand 1985; Lepescheux 1988) and the collagenous nature of this tissue has been known for years (Murray and Tanzer 1985) but new and important information has been provided only recently. The cuticle collagen consists of the longest known triple helix: 2.4μm for annelids and 1.5μm from the tube worm *Riftia*, with a terminal globular domain (Gaill et al. 1991; Gaill 1993), probably involved in oligomer formation (Gaill et al. 1995). As some of these organisms live close to hot deep sea hydrothermal vents (Gaill 1993), the thermal stability of cuticle collagen has been questionned. For *Riftia pachyptila*, the thermal stability is high (37°C), in spite of a low 4-hydroxyproline content. Recent biochemical studies have shown that the Yaa stabilizing position of 4-Hypro in the triplet Gly-Xaa-Yaa is replaced by a glycosylated threonine residue which thus appears as the major contributor to triple helix stabilization (Mann et al. 1996).

4.2.2
Nematode Cuticles

In contrast to annelid cuticle, where coarse fibers can be seen, the cuticle of nematodes is a multilayered, feltwork sheath, renewed during larval stages, and more often containing no fiber organization. In some species however, a striated material is visible (Dick and Wright 1973). This tissue is largely collagenous in nature and is formed by the mixture of a high number of different collagen polypeptides. It is assumed that the cuticle collagens are encoded by a family of at least 100 genes, many of them (at least about 60) being expressed (Cox 1992; Kramer 1994). The complete sequence of 20 of these genes has been determined in the nematode *C. elegans* and several others are known in other species (Cox et al. 1990; Van der Eycken et al. 1994; Bisoffi and Betschart 1996; Johnstone et al. 1996; Ray et al. 1996). The deduced proteins are composed of three to five short (27 to 66 residue) triple helical domains separated by non-triple helical sequences (2 to 21 residues, the longest containing 2 cysteine residues), and flanked by a short (9 to 62 residue) C-terminal domain, containing 2 cysteines and a longer (91 to 151 residue) N-terminal domain, containing three cysteines close to the triple helical domain (Kramer 1994). Non-reducible cross-links appear to be made from tyrosine residues (Cox 1992; Kramer 1994), an unusual cross-linking system for collagen (van der Rest et al. 1993). Mutations in several of these cuticle collagen genes cause alterations in the morphology and physiology of the worms (see Kramer 1994).

4.2.3
Invertebrate Exoskeletons

In the phylum Porifera, most, if not all, species are attached to their support by a collagenous cement, termed spongin. Especially in the class Demospongiae, this basal layer often forms a tree-like skeleton which can extend and constitute the actual structure supporting the animal. The best example is the bath sponge which is made of only this cleaned skeleton. In this case, it is mostly organic collagenous material. In other species, it can be mixed with sand grains or siliceous discrete elements (the spicules) secreted by special sponge cells (Garrone 1978). In some sponges, i.e. the fresh-water sponges, the basal epithelium completely surrounds the skeleton, while in others it is discontinuous inside the animal, leaving the skeleton directly in contact with the wandering sponge cells. This kind of collagenous structure has a few representatives in the invertebrates, the best of them being calcium carbonate mineralized structures found in some cnidarians (Franc 1985; Kingsley et al. 1990).

The most recent studies have been done on sponges and especially in fresh-water sponges; they are confirmed by biochemical studies of marine sponges

(Shimizu and Yoshizato 1993). Genomic and cDNA studies reveal that spongin contains a short-chain collagenous component, coded by a gene family (Exposito et al. 1990, 1991). Two genes, *COLNF13* and *COLNF8*, have been characterized. They code for two homologous chains. The chains have two collagenous domains of 171 and 66 residues, separated by a non-collagenous domain of 13 residues containing two cysteines (another cysteine being at the limit of the shortest collagenous domain). A 156 (COLNF13) or 155 (COLNF8) residue-long C-terminal domain, containing nine cysteines, appears duplicated and has homologies with the C-terminal domain of type IV collagen chains. The N-terminal domain is 51 (COLNF13) or 102 (COLNF8) residues long and contains two (COLNF13) or three (COLNF8) cysteines. This domain is hydrophobic and might be transmembranar (Lu, unpubl. data). The comparison of this sponge collagen with other collagens indicates: (1) a general feature comparable with the nematode cuticle collagens, made with several collagenous domains separated with non collagenous interruptions, but also with the vertebrate FACIT collagens having the same gross outline; (2) a C-terminal domain which is comparable with the type IV collagen NC1 domain (duplication and sequence homologies); and (3) a striking similarity to vertebrate type XIII collagen: the same length of two collagenous domains (66 and 172 residues for type XIII), and a posssible N-terminal membrane insertion. In conclusion, this sponge collagen has many of the characteristics of several non-fibrillar collagens and might be the ancestral collagen. The non-fibrillar collagen character is even more enhanced by genes organized so that exon-intron borders are not made by intact codons (Exposito et al. 1991). Possibly distinct intronless collagen genes have been described in a marine sponge (Aho et al. 1993); they could encode a spongin-like protein family.

5
Conclusion: From a Sticky Membrane Protein to an Extracellular Matrix Component

From these data, a speculative scheme of collagen evolution can be attempted. The first collagenous protein that might have appeared when unicellular organisms, perhaps colonial ones, became multicellular is most likely to have been a membrane-bound collagen similar to spongin. In contrast to polysaccharide-rich mucoid substances forming the intercellular cement of unicellular colonies, a collagenous protein is more able to make highly ordered polymers more resistant to enzyme degradation (due to the triple helical domains). The central question is to know whether this collagen appeared first as a single kind of molecule or if it appeared simultaneously with fibrillar collagen. In other words, was there only an external collagen, or an external plus an internal one? The only clear fact is that the two collagens were present very early, as is demonstrated by their presence in sponges.

Fibrillar collagen has evolved linearly, giving rise to similar α chains, leading to the five fibrillar collagens known in vertebrates. Some diversifications probably form the cuticular collagen of annelids and the bottle-brush collagen fibrils of sea urchins.

The non-fibrillar collagens certainly arose from the evolution of short-chain sponge collagen. Soon after, basement membrane collagen chains appeared; interestingly, they quickly appear as two kinds of α chains. The conversion of a membrane-bound collagen to a collagen forming a layer underlying the cell membrane is not too surprising. The evolution toward nematode cuticle collagens or gland-secreted collagens is also understandable. The formation of FACIT collagens covering collagen fibrils can be explained due to the strong adhesive properties of this short-chain collagen.

References

Aho S, Turakainen H, Onnela ML, Boedtker H (1993) Characterization of an intronless collagen gene family in the marine sponge *Microciona prolifera*. Proc Natl Acad Sci USA 90:7288–7292

Andrikopoulos K, Liu X, Keene DR, Jaenisch R, Ramirez F (1995) Targeted mutation in the *Col5A2* gene reveals a regulatory role for type V collagen during matrix assembly. Nature [Genet] 9:31–36

Arnoux B, Mérigeau K, Saludjian P, Norris F, Norris K, Bjorn S, Olsen O, Petersen L, Ducruix A (1995) The 1.6 Å structure of Kunitz-type domain from the α3 chain of human type VI collagen. J Mol Biol 246:609–617

Bailey AJ (1985) The collagen of the echinodermata. In: Bairati A, Garrone R (eds) Biology of invertebrate and lower vertebrate collagens. Plenum Press, New York, pp 345–368

Bairati A (1985) The collagens of the mollusca. In: Bairati A, Garrone R (eds) Biology of invertebrate and lower vertebrate collagens. Plenum Press, New York, pp 277–297

Bairati A, Garrone R (eds) (1985) Biology of invertebrate and lower vertebrate collagens. Plenum Press, New York, 583pp

Bairati A, De Biasi S, Cheli F, Oggioni A (1987) The head cartilage of cephalopods. I. Architecture and ultrastructure of the extracellular matrix. Tissue Cell 19:673–685

Bairati A, Cheli F, Oggioni A, Vitellaro-Zuccarello L (1989) The head cartilage of cephalopods. II. Ultrastructure of isolated native collagen fibrils and of polymeric aggregates obtained in vitro: comparison with the cartilage of mammals. J Ultrastruct Mol Struct Res 102:132–138

Bairati A, Comazzi M, Gioria M (1995) A comparative microscopic and ultrastructural study of perichondrial tissue in cartilage of *Octopus vulgaris* (Cephalopoda, Mollusca). Tissue Cell 27:515–523

Bella J, Berman HM (1996) Crystallographic evidence for C^{α}-H--O=C hydrogen bonds in a collagen triple helix. J Mol Biol 264:734–742

Bella J, Eaton M, Broadsky B, Berman HM (1994) Crystal and molecular structure of a collagen-like peptide at 1.9 Å resolution. Science 266:75–81

Bella J, Brodsky B, Berman HM (1995) Hydration structure of a collagen peptide. Structure 3:893–906

Bisoffi M, Betschart B (1996) Identification and sequence comparison of a cuticular collagen of *Brugia pahangi*. Parasitology 113:145–155

Blumberg B, MacKrell AJ, Olson PF, Kurkinen M, Monson JM, Natzle JE, Fessler JH (1987) Basement membrane procollagen IV and its specialized carboxyl domain are conserved in *Drosophila*, mouse and human. J Biol Chem 262:5947–5959

Bork P (1992) The modular architecture of vertebrate collagens. FEBS Lett 307:49–54

Bork P, Downing KA, Kieffer B, Campbell ID (1996) Structure and distribution of modules in extracellular proteins. Q Rev Biophys 29:119–167

Boute N, Exposito JY, Boury-Esnault N, Vacelet J, Noro N, Miyazaki K, Yoshizato, Garrone R (1996) Type IV collagen in sponges, the missing link in basement membrane ubiquity. Biol Cell 88:37–44

Brand DD, Blanquet RS, Phelan MA (1993) Collagenaceous, thiol-containing proteins of the cnidarian nematocyst: a comparison of the chemistry and protein distribution patterns in two types of cnidae. Comp Biochem Physiol [B] 106:115–124

Brodsky B, Shah N (1995) The triple-helix motif in proteins. FASEB J 9:1537–1546

Brodsky B, Ramshaw JAM (1997) The collagen triple-helix structure. Matrix Biol 15:545–554

Brown JC, Timpl R (1995) The collagen superfamily. Int Arch Allergy Immunol 107:484–490

Bruns RR (1984) Beaded filaments and long-spacing fibrils: relation to type VI collagen. J Ultrastruct Res 89:136–145

Caulagi VR, Rajan TV (1995) The structural organization of an α2(type IV) basement membrane collagen gene from the filarial nematode *Brugia malayi*. Mol Biochem Parasitol 70:227–229

Cecchini JP, Knibiehler B, Mirre C, Le Parco Y (1987) Evidence for a type-IV-related collagen in *Drosophila melanogaster*. Evolutionary constancy of the carboxy-terminal noncollagenous domain. Eur J Biochem 165:587–593

Chareyre P, Besson MT, Fourche J, Bosquet G (1996) Identification of a *Bombyx* collagenous protein with multiple short domains of Gly-Xaa-Yaa repeats. cDNA characterization and regulation of expression. Insect Biochem Mol Biol 26:677–685

Chen Q, Gibney E, Fitch JM, Linsenmayer C, Schmid TM, Linsenmayer TF (1990) Long-range movement and fibril association of type X collagen within embryonic cartilage matrix. Proc Natl Acad Sci USA 87:8046–8050

Christiano AM, Hoffman GG, Chung-Honet LC, Lee S, Cheng W, Uitto J, Greenspan DS (1994) Structural organization of the human type VII collagen gene (COL7A1), composed of more exons than any previously characterized gene. Genomics 21:169–179

Cox GN (1992) Molecular and biochemical aspects of nematode collagens. J Parasitol 78:1–15

Cox GN, Shamansky LM, Boisvenue RJ (1990) *Haemonchus contortus*: evidence that the 3A3 collagen gene is a member of an evolutionarily conserved family of nematode cuticle collagens. Exp Parasitol 70:175–185

Cronshaw AD, Fothergill-Gilmore LA, Hulmes DSJ (1995) The proteolytic processing site of the precursor of lysyl oxidase. Biochem J 306:279–284

D'Alessio M, Ramirez F, Suzuki H, Solursh M, Gambino R (1990) Cloning of a fibrillar collagen gene expressed in the mesenchymal cells of the developing sea urchin embryo. J Biol Chem 265:7050–7054

Dick TA, Wright KA (1973) The ultrastructure of the cuticle of the nematode *Syphacia obvelata*. Can J Zool 51:187–196

Diehl-Seifert B, Kurelec B, Zahn RK, Dorn A, Jericevic B, Uhlenbruck G, Müller WEG (1985) Attachment of sponge cells to collagen substrata: effect of a collagen assembly factor. J Cell Sci 79:271–285

Duvert M, Salat C (1990) Ultrastructural and cytochemical studies on the connective tissue of chaetognaths. Tissue Cell 22:865–878

Engel J (1991) Common structural motifs in proteins of the extracellular matrix. Curr Opin Cell Biol 3:779–785

Erlinger R, Welsh U, Scott JE (1993) Ultrastructural and biochemical observations on proteoglycans and collagen in the mutable connective tissue of the feather star *Antedon bifida* (Echinodermata, Crinoidea). J Anat 183:1–11

Exposito JY, Garrone R (1990) Characterization of a fibrillar collagen gene in sponges reveals the early evolutionary appearance of two collagen gene families. Proc Natl Acad Sci USA 87:6669–6673

Exposito JY, Ouazana R, Garrone R (1990) Cloning and sequencing of a Porifera partial cDNA coding for a short-chain collagen. Eur J Biochem 190:401–406

Exposito JY, Le Guellec D, Lu Q, Garrone R (1991) Short chain collagens in sponges are encoded by a family of closely related genes. J Biol Chem 266:21923–21928

Exposito JY, D'Alessio M, Ramirez F (1992a) Novel amino-terminal propeptide configuration in a fibrillar procollagen undergoing alternative splicing. J Biol Chem 267:17404–17408

Exposito JY, D'Alessio M, Solursh M, Ramirez F (1992b) Sea urchin collagen evolutionarily homologous to vertebrate pro-α2(I) collagen. J Biol Chem 267:15559–15562

Exposito JY, D'Alessio M, Di Liberto M, Ramirez F (1993a) Complete primary structure of a sea urchin type IV collagen chain and analysis of the 5' end of its gene. J Biol Chem 268:5249–5254

Exposito JY, van der Rest M, Garrone R (1993b) The complete intron/exon structure of *Ephydatia mülleri* fibrillar collagen gene suggests a mechanism for the evolution of an ancestral gene module. J Mol Evol 37:254–259

Exposito JY, Suzuki H, Geourjon C, Garrone R, Solursh M, Ramirez F (1994) Identification of a cell lineage-specific gene coding for a sea urchin α2(IV)-like collagen chain. J Biol Chem 269:13167–13171

Exposito JY, Boute N, Deleage G, Garrone R (1995) Characterization of two genes coding for a similar four-cysteine motif of the amino-terminal propeptide of a sea urchin fibrillar collagen. Eur J Biochem 234:59–65

Fessler JH, Fessler LI (1989) *Drosophila* extracellular matrix. Annu Rev Cell Biol 5:309–339

Fessler LI, Condic ML, Nelson RE, Fessler JH, Fristrom JW (1993) Site-specific cleavage of basement membrane collagen IV during *Drosophila* metamorphosis. Development 117:1061–1069

Franc S (1985) Collagen of coelenterates. In: Bairati A, Garrone R (eds) Biology of invertebrate and lower vertebrate collagens. Plenum Press, New York, pp 197–210

Francois J (1985) The collagen of the Arthropoda. In: Bairati A, Garrone R (eds) Biology of invertebrate and lower vertebrate collagens. Plenum Press, New York, pp 345–368

Gaill F (1993) Aspects of life development at deep sea hydrothermal vents. FASEB J 7:558–565

Gaill F, Bouligand Y (1985) Long pitch helices in invertebrate collagens. In: Bairati A, Garrone R (eds) Biology of invertebrate and lower vertebrate collagens. Plenum Press, New York, pp 267–274

Gaill F, Wiedemann H, Mann K, Kühn K, Timpl R, Engel J (1991) Molecular characterization of cuticle and interstitial collagens from worms collected at deep sea hydrothermal vents. J Mol Biol 221:209–223

Gaill F, Mann K, Wiedemann H, Engel J, Timpl R (1995) Structural comparison of cuticle and interstitial collagens from annelids living in shallow sea-water and at deep-sea hydrothermal vents. J Mol Biol 246:284–294

Garrone R (1978) Phylogenesis of connective tissue. Karger, Basel

Garrone R (1984) Formation and involvement of extracellular matrix in the development of sponges, a primitive multicellular system. In: Trelstad RL (ed) The role of extracellular matrix in development. Liss, New York, pp 461–477

Garrone R (1985) The collagen of the Porifera. In: Bairati A, Garrone R (eds) Biology of invertebrate and lower vertebrate collagens. Plenum Press, New York, pp 157–175

Garrone R, Huc A, Junqua S (1975) Fine structure and physicochemical studies on the collagen of the marine sponge *Chondrosia reniformis* Nardo. J Ultrastruct Res 52:261–275

Gramzow M, Schröder HC, Uhlenbruck G, Batel R, Müller WEG (1988) Sponge aggregation factor: identification of the specific collagen-binding site by means of a monoclonal antibody. J Histochem Cytochem 36:205–212

Greenspan DS (1993) The carboxyl-terminal half of type VII collagen, including the non-collagenous NC-2 domain and intron/exon organization of the corresponding region of the COL7A1 gene. Hum Mol Gen 2:273–278

Gross J, Sokal Z, Rougvie M (1956) Structural and chemical studies on the connective tissue of marine sponges. J Histochem Cytochem 4:227–246

Guo X, Kramer JM (1989) The two *Caenorhabditis elegans* basement membrane (type IV) collagen genes are located on separate chromosomes. J Biol Chem 264:17574–17582

Guo X, Johnson JJ, Kramer JM (1991) Embryonic lethality caused by mutations in basement membrane collagen of *C. Elegans*. Nature 349:707–709

Hermans CO (1970) The periodicity of collagen in the brain sheath of a polychaete. J Ultrastruct Res 30:255–261

Holstein TW, Benoit M, Herder GV, Wanner G, David CN, Gaub HE (1994) Fibrous mini-collagens in *Hydra* nematocysts. Science 265:402–404

Hopkinson SB, Baker SE, Jones JCR (1995) Molecular genetic studies of a human epidermal autoantigen (the 190-kDa bullous pemphigoid antigen/BP 180): identification of functionally important sequences within the BP 180 molecule and evidence for an interaction between BP 180 and α6 integrin. J Cell Biol 130:117–125

Hudson BG, Reeders SY, Tryggvason K (1993) Type IV collagen: structure, gene organization and role in human diseases. J Biol Chem 268:26033–26036

Humbert-David N, Garrone R (1993) A six-armed tenascin-like protein extracted from the Porifera *Oscarella tuberculata* (Homosclerophorida). Eur J Biochem 216:255–260

Hunt S (1985) The selachian egg case collagen. In: Bairati A, Garrone R (eds) Biology of invertebrate and lower vertebrate collagens. Plenum Press, New York, pp 409–434

Johnstone IL, Shafi Y, Majeed A, Barry JD (1996) Cuticular collagen genes from the parasitic nematode *Ostertagia circumcincta*. Mol Biochem Parasitol 80:103–112

Kellokumpu S, Sormunen R, Heikkinen J, Myllylä R (1994) Lysyl hydroxylase, a collagen processing enzyme, exemplifies a novel class of luminally-oriented peripheral membrane proteins in the endoplasmic reticulum. J Biol Chem 269:30524–30529

Kellokumpu S, Suoka M, Risteli L, Myllylä R (1997) Protein disulfide isomerase and newly synthesized procollagen chains form higher-order structures in the lumen of the endoplasmic reticulum. J Biol Chem 272:2770–2777

Kimura S, Takema Y, Kubota M (1981) Octopus skin collagen. J Biol Chem 256:13230–13234

Kimura S, Omura Y, Ishida M, Shirai H (1993) Molecular characterization of fibrillar collagen from the body wall of starfish *Asterias amurensis*. Comp Biochem Physiol [B] 104:663–668

Kingsley RJ, Tsuzaki M, Watabe N, Mechanic GL (1990) Collagen in the spicule organic matrix of the gorgonian *Leptogorgia virgulata*. Biol Bull 179:207–213

Kivirikko KI, Myllylä R, Pihlajaniemi T (1989) Protein hydroxylation: prolyl 4-hydroxylase, an enzyme with four cosubstrates and a multifunctional subunit. FASEB J 3:1609–1617

Knight DP, Feng D, Stewart M (1996) Structure and function of the selachian egg case. Biol Rev 71:81–111

Koivu J, Myllylä R, Helaakoski T, Pihlajaniemi T, Tasanen K, Kivirikko KI (1987) A single polypeptide acts both as the β subunit of prolyl 4-hydroxylase and as a protein disulfide isomerase. J Biol Chem 262:6447–6449

Kong RYC, Kwan KM, Lau ET, Thomas JT, Boot-Handford RP, Grant ME, Cheah KSE (1993) Intron-exon structure, alternative use of promoter and expression of the mouse collagen X gene, Col10a-1. Eur J Biochem 213:99–111

Kramer JM (1994) Structures and functions of collagens in *Caenorhabditis elegans*. FASEB J 8:329–336

Kühn K (1994) Basement membrane (type IV) collagen. Matrix Biol 14:439–445

Kurz EM, Holstein TW, Petri BM, Engel J, David CN (1991) Mini-collagens in *Hydra* nematocytes. J Cell Biol 115:1159–1169

Kwan APL, Cummings CE, Chapman JA, Grant ME (1991) Macromolecular organization of chicken type X collagen in vitro. J Cell Biol 114:597–604

Ledger PW (1974) Types of collagen fibres in the calcareous sponges *Sycon* and *Leucandra*. Tissue Cell 6:385–389

Leivo I, Tani T, Laitinen L, Bruns R, Kivilaakso E, Lehto V, Burgeson RE, Virtanen I (1996) Anchoring complex components laminin-5 and type VII collagen in intestine: association with migrating and differentiating enterocytes. J Histochem Cytochem 44:1267–1277

Lepescheux L (1988) Spatial organization of collagen in annelid cuticle: order and defects. Biol Cell 62:17–31

Lethias C, Exposito JY, Garrone R (1997) Collagen fibrillogenesis during sea urchin development. Retention of SURF motifs from the N-propeptide of the 2α chain in mature fibrils. Eur J Biochem 245:434–440

Li K, Sawamura D, Giudice GJ, Diaz LA, Mattei MG, Chu ML, Uitto J (1991) Genomic organization of collagenous domains and chromosomal assignment of human 180 kDa bullous pemphigoid antigen-2, a novel collagen of stratified squamous epithclium. J Biol Chem 266:24064–24069

Li K, Tamai K, Tan EM, Uitto J (1993) Cloning of type XVII collagen. J Biol Chem 268:8825–8834

Li Y, Lacerda DA, Warman ML, Beier DR, Yoshioka H, Ninomiya Y, Oxford JT, Morris NP, Andrikopoulos K, Ramirez F, Wardell BB, Lifferth GD, Teuscher C, Woodward SR, Taylor BA, Seegmiller RE, Olsen BR (1995) A fibrillar collagen gene, Col11a1, is essential for skeletal morphogenesis. Cell 80:423–430

Lu Valle P, Iwamoto M, Fanning P, Pacifici M, Olsen BR (1993) Multiple negative elements in a gene that codes for an extracellular matrix protein, collagen X, restrict expression to hypertrophic chondrocytes. J Cell Biol 121:1173–1179

Mann K, Gaill F, Timpl R (1992) Amino-acid sequence and cell-adhesion activity of a fibril-forming collagen from the tube worm *Riftia pachyptila* living at deep sea hydrothermal vents. Eur J Biochem 210:839–847

Mann K, Mechling DE, Bächinger HP, Eckerskorn C, Gaill F, Timpl R (1996) Glycosylated threonine but not 4-hydroxyproline dominates the triple helix stabilizing positions in the sequence of a hydrothermal vent worm cuticle collagen. J Mol Biol 261:255–266

MacBeath JRE, Kielty CM, Shuttleworth CA (1996) The type VIII collagen is a product of vascular smooth-muscle cells in development and disease. Biochem J 319:993–998

Mazzorana M, Gruffat H, Sergeant A, van der Rest M (1993) Mechanism of collagen trimer formation. J Biol Chem 268:3029–3032

Mizuta S, Yoshinaka R, Sato M, Sakaguchi M (1994) Characterization of collagen in muscle of several crustacean species. Comp Biochem Physiol [B] 107:365–370

Muragaki Y, Jacenko O, Apte S, Mattei MG, Ninomiya Y, Olsen BR (1991) The a2(VIII) collagen gene. J Biol Chem 266:7721–7727

Murray LW, Tanzer ML (1985) The collagen of Annelida. In: Bairati A, Garrone R (eds) Biology of invertebrate and lower vertebrate collagens. Plenum Press, New York, pp 243–258

Noelken ME, Wisdom BJ, Dean DC, Hung CH, Hudson BG (1986) Intestinal basement membrane of *Ascaris suum*. Molecular organization and properties of the collagen molecules. J Biol Chem 261:4706–4714

Olsen BR (1996) Collagen it takes and bone it makes. Curr Biol 6:645–647

Omura Y, Urano N, Kimura S (1996) Occurrence of fibrillar collagen with structure of (α1)2α2 in the test of sea urchin *Asthenosoma ijimai*. Comp Biochem Physiol [B] 115:63–68

Pedersen KJ (1968) Some morphological and histochemical aspects of nemertean connective tissue. Z Zellforsch 90:570–595

Pettitt J, Kingston IB (1994) Developmentally regulated alternative splicing of a nematode type IV collagen gene. Dev Biol 161:22–29

Piez KF, Gross J (1959) The amino acid composition and morphology of some invertebrate and vertebrate collagens. Biochim Biophys Acta 34:24–29

Pihlajaniemi T, Tamminen M (1990) The α1 chain of type XIII collagen consists of three collagenous and four noncollagenous domains, and its primary transcript undergoes complex alternative splicing. J Biol Chem 265:16922–16928

Prockop DJ, Kivirikko KI (1995) Collagens: molecular biology, diseases, and potentials for therapy. Annu Rev Biochem 64:403–434

Ray C, Wang TY, Hussey RS (1996) Identification and characterization of the Meloidogyne incognita *col1* cuticle collagen gene. Mol Bioch Parasitol 83:121–124

Reber-Müller S, Spissinger T, Scuchert P, Spring J, Schmid V (1995) An extracellular matrix protein of jellyfish homologous to mammalian fibrillins forms different fibrils depending on the life stage of the animal. Dev Biol 169:662–672

Revenko I, Sommer F, Tran Minh D, Garrone R, Franc JM (1994) Atomic force microscopy study of collagen fibre structure. Biol Cell 80:67–69

Sakai L, Keene DR, Morris NP, Burgeson RE (1986) Type VII collagen is a major structural component of anchoring fibrils. J Cell Biol 103:1577–1586

Sarras MP, Madden ME, Zhang X, Gunwar S, Huff JK, Hudson BG (1991) Extracellular matrix (mesoglea) of *Hydra* vulgaris. I. Isolation and characterization. Dev Biol 148:481–494

Sawada H, Konomi H, Hirosawa K (1990) Characterization of the collagen in the hexagonal lattice of Descemet's membrane: its relation to type VIII collagen. J Cell Biol 110:219–227

Schmut O, Roll P, Reich ME (1980) Biochemical and electronmicroscopic investigations on *Helix pomatia* collagen. Z Naturforsch 35c:376–379

Shimizu K, Yoshizato K (1993) Involvement of collagen synthesis in tissue reconstitution by dissociated sponge cells. Dev Growth Differ 35:293–300

Sibley MH, Johnson JJ, Mello CC, Kramer JM (1993) Genetic identification, sequence and alternative splicing of *Caenorhabditis elegans* $\alpha2$(IV) collagen gene. J Cell Biol 123:255–263

Sibley MH, Graham PL, von Mende N, Kramer JM (1994) Mutations in the $\alpha2$(IV) basement membrane collagen gene of *Caenorhabditis elegans* produce phenotypes of differing severities. EMBO J 13:3278–3285

Sicot FX, Exposito JY, Masselot M, Garrone R, Deutsch J, Gaill F (1997) Cloning of an annelid fibrillar-collagen gene and phylogenetic analysis of vertebrate and invertebrate collagens. Eur J Biochem 246:50–58

Takahara K, Hoffman GG, Greenspan DS (1995) Complete structural organization of the human $\alpha1$(V) collagen gene (*COL5A1*): divergence from the conserved organization of other characterized fibrillar collagen genes. Genomics 29:588–597

Thurmond FA, Trotter JA (1994) Native collagen fibrils from echinoderms are molecularly bipolar. J Mol Biol 235:73–79

Tillet E, Franc JM, Franc S, Garrone R (1996) The evolution of fibrillar collagens: a sea-pen collagen shares common features with vertebrate type V collagen. Comp Biochem Physiol [B] 113:239–246

Tillet-Barret E, Franc JM, Franc S, Garrone R (1992) Characterization of heterotrimeric collagen molecules in a sea-pen (Cnidaria, Octocorallia). Eur J Biochem 203:179–184

Timpl R (1996) Macromolecular organization of basement membranes. Curr Opin Cell Biol 8:618–624

Trotter JA, Koob TJ (1989) Collagen and proteoglycan in a sea urchin ligament with mutable mechanical properties. Cell Tissue Res 258:527–539

Trotter JA, Koob TJ (1994) Biochemical characterization of fibrillar collagen from the mutable spine ligament of the sea-urchin *Eucidaris tribuloides*. Comp Biochem Physiol [B] 107:125–134

Trotter JA, Thurmond FA, Koob TJ (1994) Molecular structure and functional morphology of echinoderm collagen fibrils. Cell Tissue Res 275:451–458

Trotter JA, Lyons-Levy G, Thurmond FA, Koob TJ (1995) Covalent composition of collagen fibrils from the dermis of the sea cucumber, *Cucumaria frondosa*, a tissue with mutable mechanical properties. Comp Biochem Physiol [A] 112:463–478

Väisänen T, Hägg P, Huhtala P, Rehn M, Pihlajaniemi T (1996) Type XIII collagen is a focal adhesion protein with possible adhesive function. Matrix Biol 15:160

Van der Eycken W, de Almeida Engler J, Van Montagu M, Gheysen G (1994) Identification and analysis of a cuticular collagen-encoding gene from the plant-parasitic nematode *Meloidogyne incognita*. Gene 151:237–242

Van der Rest M, Garrone R (1990) Collagen as multidomain proteins. Biochimie 72:473–484

Van der Rest M, Garrone R (1991) The collagen family of proteins. FASEB J 5:2814–2823

Van der Rest M, Garrone R, Herbage D (1993) Collagen: a family of proteins with many facets. Adv Mol Cell Biol 6:1–67

Vitellaro-Zuccarello L, Cheli F, Cetta G (1985) The interstitial collagen of *Lumbricus* sp. (Annelida). In: Bairati A, Garrone R (eds) Biology of invertebrate and lower vertebrate collagens. Plenum Press, New York, pp 259–265

Vuoristo MM, Pihlajamaa T, Vandenberg P, Prockop DJ, Ala-Kokko L (1995) The human *COL11A2* gene structure indicates that the gene has not evolved with the genes for the major fibrillar collagens. J Biol Chem 270:22873–22881

Wallis GA (1993) Here today, bone tomorrow. Curr Biol 3:687–689

Weber C, Schmid V (1985) The fibrous system in the extracellular matrix of hydromedusae. Tissue Cell 17:811–822

Yasothornsrikul S, Wendy DJ, Cramer G, Kimbrell DA, Dearolf CR (1997) *Viking*: identification and characterization of a novel type IV collagen in *Drosophila*. Gene 198:17–25

Zweckstetter M, Czisch M, Mayer U, Chu ML, Zinth W, Timpl R, Holak TA (1996) Structure and multiple conformations of the Kunitz-type domain from human type VI collagen $\alpha3$(VI) chains in solution. Structure 4:195–209

Evolution of Early Metazoa: Phylogenetic Status of the Hexactinellida Within the Phylum of Porifera (Sponges)

W.E.G. Müller[1], M. Kruse[1], C. Koziol[1], J.M. Müller,[1] and S.P. Leys[2]

1
Introduction

The evolution of the Metazoa from their protozoan ancestors is one of the greatest puzzles of phylogeny (Willmer 1994; Cavalier-Smith et al. 1996). The emergence of multicellular animals has been explained by two major theories: the syncytial theory (Hadzi 1963) – origin from a multinucleated ciliate – and the colonial theory (Haeckel 1868) – origin from a colonial flagellate – both of which assume a di(poly)phyletic origin of the Metazoa. Numerous attempts to resolve whether the Metazoa are of mono- or polyphyletic origin have sought evidence from a wide variety of developmental and morphological data such as body symmetry, type of development (protostome vs. deuterostome), type of body cavities (coelomates vs. pseudocoelomates), cleavage patterns (determinate vs. indeterminate, spiral vs. radial), origin of photoreceptors, evolution of surface cilitation, etc. (see Morris 1993).

More recently, molecular data has been applied in the hope of solving the question of metazoan phylogeny definitively. Early molecular phylogenetic studies using nucleotide sequence data from rRNA, but excluding the lowest metazoan phylum, the Porifera (sponges), concluded that multicellular animals evolved only once (Field et al. 1988; Lake 1990). However, if sequences from sponges, derived from 18S (Field et al. 1988) or 28S rRNA (Christen et al. 1991), were included, the analysis indicated that the Radiata (including Porifera, Placozoa, Cnidaria and Ctenophora) and the Bilateria (other animal phyla) originated separately from different protozoan ancestors, once again promoting the theory that the Metazoa are diphyletic. Later it was concluded that 18S rRNA sequence analysis was not suitable for resolving deep branching points in the phylogenetic trees, such as the positioning of the phylum Porifera within the Metazoa (Rodrigo et al. 1994).

It has been suggested that very few, perhaps only two, developmental strategies could have allowed the transition from protists to metazoans: aggregation of mitotically-related or unrelated cells, or division of formerly

[1]Institut für Physiologische Chemie, Abteilung Angewandte Molekularbiologie, Johannes Gutenberg-Universität, Duesbergweg 6, 55099 Mainz, Germany
[2]Department of Biology, University of Victoria, P.O. Box 1700, Victoria, B.C., V8W 2Y2, Canada

Progress in Molecular and Subcellular Biology, Vol. 21
W.E.G. Müller (Ed.)
© Springer-Verlag Berlin Heidelberg 1998

multinucleate cells which are formed by incomplete division of the cytoplasm (Willmer 1994). In both cases the unicellular ancestors of the Metazoa must have acquired the ability to interact firstly with other cells and subsequently with the extracellular matrix. Our approach to examine the question of the phylogeny of metazoans was to isolate and analyse genes coding for proteins required for multicellularity, such as cell adhesion molecules and receptors from sponges (reviewed in Müller 1997b). These data clearly indicated that the Porifera are metazoans, and that all Metazoa are of monophyletic origin (Müller et al. 1994; Gamulin et al. 1994; Müller 1995, 1997a).

However, one major question that remained was the phylogenetic position of the class Hexactinellida within the phylum Porifera.

2
Hexactinellida

Sponges in the class Hexactinellida are syncytial rather than cellular. Approximately 75% of the tissue forms a multinucleate syncytium; the remaining tissue consists of uninucleate "cells" which are connected to the syncytium by perforate (aqueous) plugged junctions (Mackie 1981; Mackie and Singla 1983; Leys 1996). Although the syncytial nature of the Hexactinellida was known from histological studies conducted on poorly preserved tissue in the late 1800s, it was not confirmed until recently when improved preservation techniques were developed for tissue to be examined by electron microscopy (e.g. Mackie and Singla 1983; Mehl and Reiswig 1991; Boury-Esnault and de Vos 1988), and live tissue cultures were studied by video microscopy (Leys 1995, 1997). Work with live tissue has shown that dissociated tissue from hexactinellids aggregates by fusion to form a giant syncytium. Within the syncytium the cytoplasm is continuously transported along bundles of microtubules for distances greater than several centimetres (Leys 1995, 1996, 1997). Although the mechanism of transport appears to be similar to that found intracellularly in demosponges and the Calcarea (see Wachtmann et al. 1990), the distances organelles are transported are much greater, and in hexactinellids, even nuclei are transported (Leys 1995, 1997).

A fundamental difference between hexactinellids and cellular sponges, which stems from the difference in tissue structure, is that the hexactinellid syncytium allows the rapid propagation of electrical events which co-ordinate the shutdown of flagellar beating (Leys and Mackie 1997). Cellular sponges are not known to possess aqueous intercellular junctions which would allow rapid communication between cells, and consequently these sponges are only capable of slow (several minutes in duration), localised contractions, which are likely to be mechanical in origin. Impulses in *Rhabdocalyptus dawsoni*, in contrast, travel at ~0.2 cm/s which approaches the velocity of propagated events in other invertebrate epithelia (Mackie 1965; Anderson 1980). Recent data suggests that the propagated impulse in *R. dawsoni* may be by way of a calcium spike (Mackie et al., in prep.), which would be similar to the mechanism of impulse conduction in some protists. However, although sodium channels first appear in the Cnidaria, a thorough search for them in the Porifera has not yet been conducted, and more detailed experiments are required to rule out conduction by way of a sodium current in hexactinellids. Furthermore, since epithelial conduction has evolved independently many times in the Metazoa, this in itself is not an indication of either a primitive or advanced condition.

3
Problem of Classification of Hexactinellida

There have been several attempts to separate the Hexactinellida from other sponges at the subphylum level, based on their remarkably different tissue structures. The most recent proposal, which resulted from extensive ultrastructural studies and the discovery that, unlike cellular sponges, *R. dawsoni* could propagate behaviourally meaningful events, suggested dividing the Porifera into the subphylum Cellularia, which includes the classes Demospongiae and Calcarea, and the subphylum Symplasma, with only one class, the Hexactinellida (Reiswig and Mackie 1983). The fundamental structural difference between these sponges raises the question as to whether the ancestors of the Metazoa in general, and the Porifera in particular, were colonial flagellates (Hyman 1940) or syncytial ciliates (Hanson 1977). Two alternative hypotheses have been proposed to explain the relationships between the major sponge classes. One groups the Porifera into the adelphotaxa Hexactinellida and Demospongiae/Calcarea based on the gross difference in tissue structure and on differences in the structure of the flagella, whose beating generates the feeding current through sponges (Mehl and Reiswig 1991). The other hypothesis assumes that the Demospongiae are more closely related to the Hexactinellida, based on presumed larval similarities (Böger 1988). There is, however, very little information available on hexactinellid larvae.

4
A Rational Solution: The Deduced Amino Acid Sequence of Protein Kinase C

Prior to the present study, no cDNA coding for a protein had been cloned from the Hexactinellida, hence no phylogenetic analysis based on such molecules was possible. We have now isolated a cDNA encoding one key enzyme involved in signal transduction pathways, a protein-serine/threonine kinase (Nishizuka 1992), from the hexactinellid sponge *R. dawsoni*. The analysis of the deduced amino acid (aa) sequence of this protein kinase C (PKC) belonging to the subfamily cPKC, using both the catalytic domain and the regulatory part of the enzyme, composed of the pseudosubstrate site, the Cys/His effector binding domains (zinc fingers) and the C2 domain (Ca^{2+} specific domain), indicates that the Hexactinellida diverged earlier than either the

Demospongiae or the Calcarea, from a common ancestor. In addition, it is shown that according to this analysis, the cPKC from the Calcarea shares a common ancestor with protostome and deuterostome cPKCs.

5
The Sponge cPKC Sequences

5.1
cDNAs

The cDNAs encoding cPKCs have been cloned from the following sponge species: *Geodia cydonium* [Jameson] (Metazoa; Porifera, Demospongiae, Tetractinomorpha, Astrophorida, Geodiidae), *Suberites domuncula* [Olivi] (Metazoa; Porifera, Demospongiae, Tetractinomorpha, Hadromerida, Suberitidae) and *Sycon raphanus* [Schmidt] (Metazoa; Porifera, Calcarea, Calcarona, Leucosoleniida, Sycettidae), all collected near Rovinj (Croatia), and *Rhabdocalyptus dawsoni* [Lambe] (Metazoa; Porifera; Hexactinellida; Hexasterophora; Lyssacinosida; Rossellidae), which was collected from Saanich Inlet and Barkley Sound, British Columbia (Canada) by SCUBA diving. Since *R. dawsoni* hosts a number of other invertebrates on its outer spicules, cleaned sponge tissue was dissociated through 150 μm Nitex mesh and allowed to reaggregate over 12h. Aggregates were transferred to 0.45 μm Millipore filtered seawater three times in 2 days prior to freezing in liquid nitrogen to reduce the likelihood of contamination by other organisms.

The cloning strategies for obtaining the cDNAs will be outlined separately. The nucleotide (nt) sequences as well as the deduced aa sequence from the sponges have been deposited in the EMBL data base: *S. domuncula* cPKC [*CPKCSD*] Y10529, *S. raphanus* cPKC [*CPKCSR*] Y10530, *G. cydonium* cPKC [*CPKCGC*] X87683 and *R. dawsoni* cPKC [*CPKC_RD*] Y13229. The detailed description of the respective aa sequences are given elsewhere: they are named: *S. domuncula* cPKC cPKC_SD, *S. raphanus* cPKC cPKC_SR, *G. cydonium* cPKC cPKC_GC and *R. dawsoni* cPKC cPKC_RD.

5.2
Cloning of the cDNA Encoding a "Conventional" PKC from *R. dawsoni*

The cDNA encoding a PKC from the hexactinellid sponge *R. dawsoni* was isolated and characterised. The clone is 2251-bp-long and has an open reading frame of 2091 nt. The predicted translation product of 697 aa of CPKC_RD (Fig. 1A) has a size of M_r 79377. The following sites, which are characteristic for the subfamily of the "Conventional" PKC (cPKC) are present in CPKC_RD (Fig. 1A; Stabel and Parker 1991; Hardie and Hanks 1995): the pseudosubstrate prototype (aa 17–33), two zinc fingers (47–83 and 114–152), the C2 domain, which provides the Ca^{2+}-dependence (187–201), and the catalytic domain (343–635). A scheme of the putative *R. dawsoni* cPKC is shown in Fig. 1B.

A

```
MAVVDPEEEISGRRVDIARRRAIREKKRVVIKDHSFEQHFFKQPTYCSHCRNFIYGVGRQGFECVQCGYVLHRHCVDLVA  80
         ~pseudosubstrate~                      ++++++++++++++++Zn-1++++++++++++++

FNCPKSVKGIAVLKKGEGKVHDFQVHSYGKFTFCDHCGQMLYGLYRQGYQCGDPDCGINVHSYCKPFVANLCGTHSNEKY 160
+++                           +++++++++++++++++Zn-2+++++++++++++++++
GRLQMNIRFNEMNERDPVIVIDIKQGKNLLPSDADGMSNPYCQVSMLDKKAKVVITHKTEVAEQSLAPIFNESFVIPFNP 240
                  ======C2=======
SKCERMSVEVMDFSGKFLGGMSFGLSELTNFKKPEDIWFTLYNDNVAKHLNNRTIIESDEAEIKRLIDQVATVRYSSSPT 320
KSLPLTQSLGACVIPGMTSDDVMVLEQYKLLGVLGKGAFGKVFMAESSDLTEEIFALKVMHKELVFDFDAVDCTMMEKLV 400
              #>> CD :::
LALEDKNKPPFLTALKASFQDQENLYMVMEFLSGGDLLFHLLKSIRFTDERSKFYIAEISLAIFFLHSKEIIYRDLKPDN 480
VLLSAEGHIKLADFGLCKEGIKNGKSTTRTYCGTLDYMAPEILQRKDYDNAVDMWSLGVMLYEMLAGRTPFDARTEEMVV 560
TNIKNRAPSFPNNFSKEAIKLITVLLKKVPKERLGFDPVDGQKQFQANKYFSDIDWDMLNKQQVEPPFIPTIKGKKSTAN 640
                                                     ::: CD <<#
FDKEFTQASNKITKSASRNTIMQHEDFDGFEYTSPDLRQAMRIKREATEQIENEATN                         697
```

B

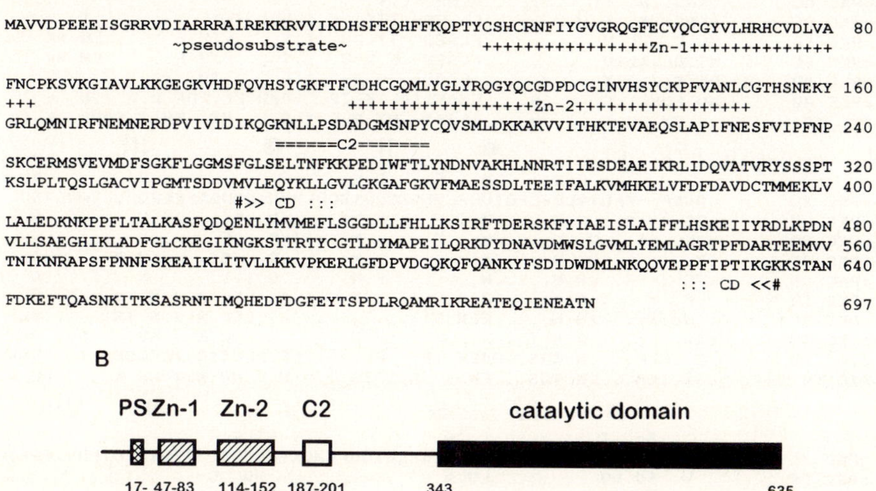

Fig. 1. A Deduced aa sequence of the cPKC from *Rhabdocalyptus dawsoni*. The following sites are marked: pseudosubstrate sequence (~~~), two zinc fingers (*Zn-1* and *Zn-2*; +++), *C2* domain (===). The boundaries of the catalytic domain (*CD*) are shown (#<< >>#). **B** Scheme of the *R. dawsoni* cPKC; the two conserved parts of the kinase (1) the regulatory region, including the pseudosubstrate site, the two zinc fingers and the C2 domain and (2) the catalytic domain, are shown; the *numbers* indicate the borders of the respective regions

5.3
Phylogenetic Position of *R. dawsoni*: Analysis of the Catalytic Domain

Previous extensive comparative work with sequences of cPKC from Bacteria, Protozoa, Metazoa, and Planta (Hanks et al. 1988; Hunter 1991; Hardie and Hanks 1995) has shown that both regions of the kinase – the highly conserved part of the catalytic domain and the *N*-terminal end with the highly conserved regulatory blocks, which are interspersed by less conserved sequences – can be used separately for phylogenetic analyses.

The alignment of the catalytic domains, shown in Fig. 2, includes sequences for cPKC Ser/Thr-protein kinases from metazoans and from the yeast *Saccharomyces cerevisiae*. The metazoan sequences comprise sponge sequences isolated from (1) the demosponges *G. cydonium* and *S. domuncula*, (2) the calcareous sponge *S. raphanus*, (3) the hexactinellid sponge *R. dawsoni*, and from higher metazoan phyla, from (1) the pseudocoelomate [Nematoda] *Caenorhabditis elegans*, from (2) two protostomes, *Aplysia californica* and *Drosophila melanogaster*, and from (3) two deuterostomes *Lytechinus pictus* and *Xenopus laevis*.

Sequence alignment (Fig. 2) revealed that the catalytic domain from *R. dawsoni* has highest similarity with the sequence from the demosponges

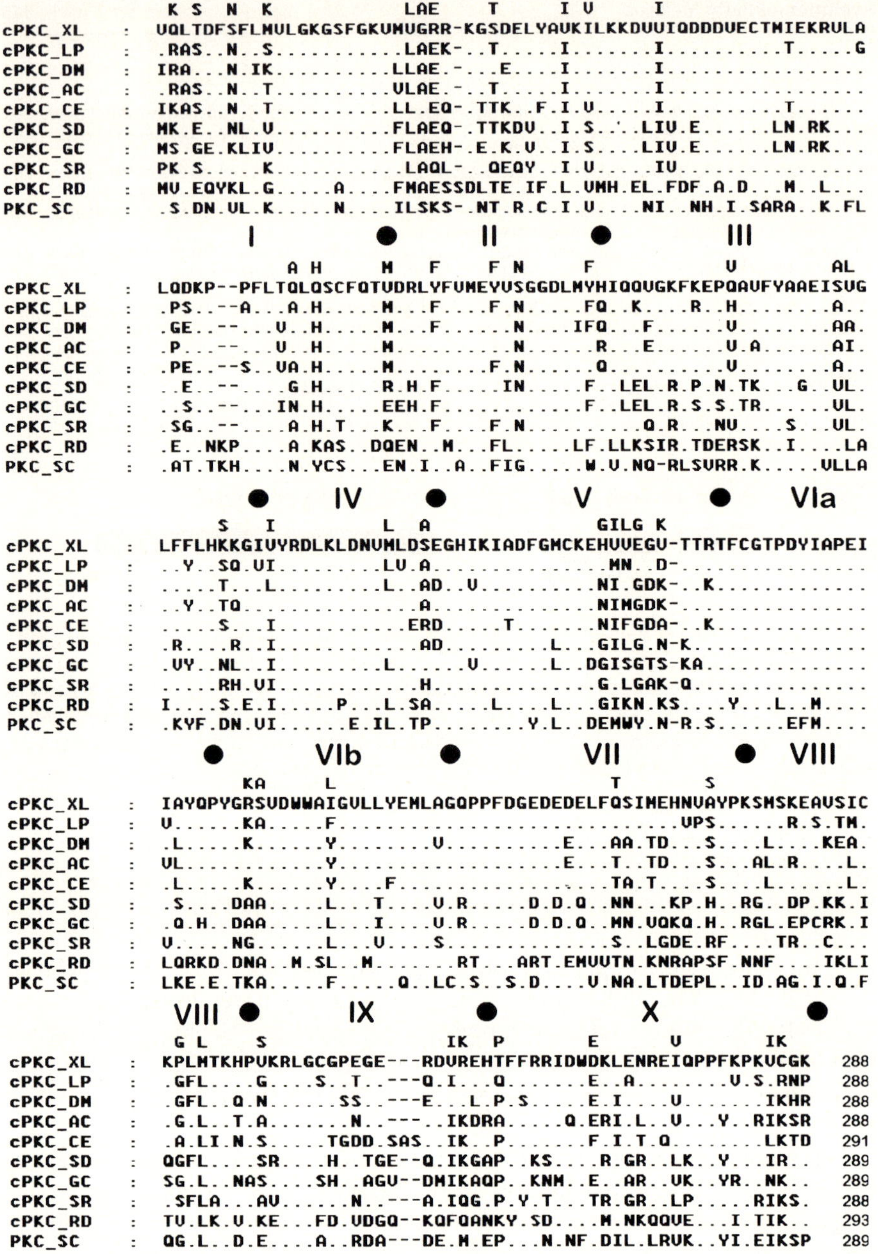

Fig. 2. Alignment of the catalytic domains of deduced PKCs from I Metazoa, cPKC from the deuterostomes *Xenopus laevis* (frog – *cPKC_XL*, accession number g104167) and *Lytechinus pictus* (sea urchin – *PKC_LP*, U02967), from the protostomes cPKC from *Drosophila melanogaster* (fruit fly – *PKC_DM*, P13678) and *Aplysia californica* (mollusc – *PKC_AC*, g228058) and from the sponges of the classes (1) Demospongiae, *Geodia cydonium* (*CPKC_GC*, X87683) and *Suberites domuncula* (*CPKC_SD*, Y10529), (2) Calcarea, *Sycon raphanus* (*CPKC_SR*, Y10530), and (3) the Hexactinellida, *R. dawsoni* (*CPKC_RD*, described here) as well as from II Yeast *Saccharomyces cerevisiae* (*PKC_SC*, P12688). Residues identical to the top sequences are indicated by a *dot*. The consensus is given in the first line above the alignment. All subdomains are shown (*I to XI*); the nomenclature follows that of Hardie and Hanks (1995). The *numbers at the end* refer to the aa residues of the respective catalytic domains

S. domuncula (49% identical aa, 70% related aa) and *G. cydonium* (48, 70%); lower similarities were found to the calcareous sponge *S. raphanus* (46, 67%) and to members of the higher metazoan phyla, *L. pictus* (45, 69%), *C. elegans* (45, 66%), *D. melanogaster* (44, 67%), *A. californica* (46, 68%) and *X. laevis* (46, 69%). Comparably little homology was found to the yeast *S. cerevisiae* (41, 63%).

Phylogenetic trees were constructed and statistical bootstrap probabilities were calculated (Figs. 3–5). The earliest offshoot was found to be the yeast

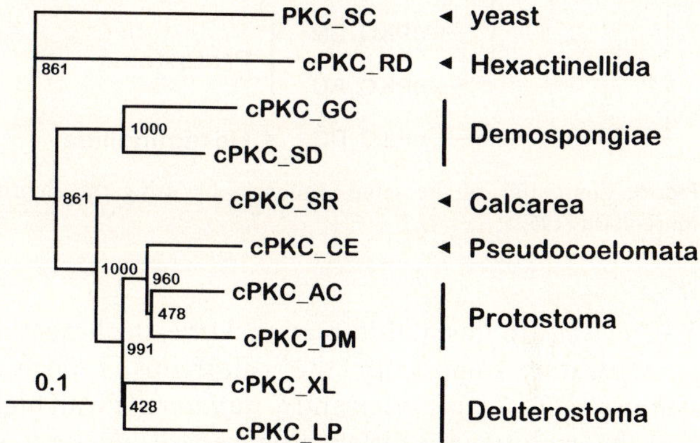

Fig. 3. Unrooted phylogenetic tree computed from the ten PKC sequences (catalytic domain). Phylogenetic trees were constructed on the basis of aa sequence alignment by using the CLUSTAL-W program (Higgins and Sharp 1988). The degree of support for internal branches was further assessed by bootstrapping (1000 bootstrap replicates) (Felsenstein 1993). The distance matrix was calculated as described (Dayhoff et al. 1978). The *scale bar* indicates an evolutionary distance of 0.1 aa substitutions per position in the sequence

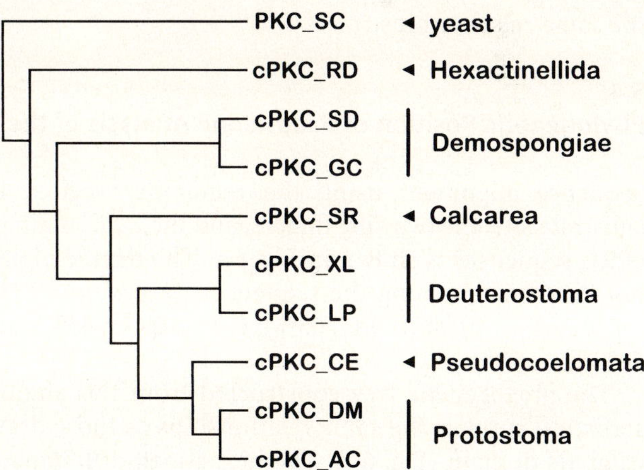

Fig. 4. Computing of the same sequences by the procedure of neighbor-joining applying the "Neighbor" program from the PHYLIP package PROTPARS. (Protein-Parsinomy; Felsenstein 1993)

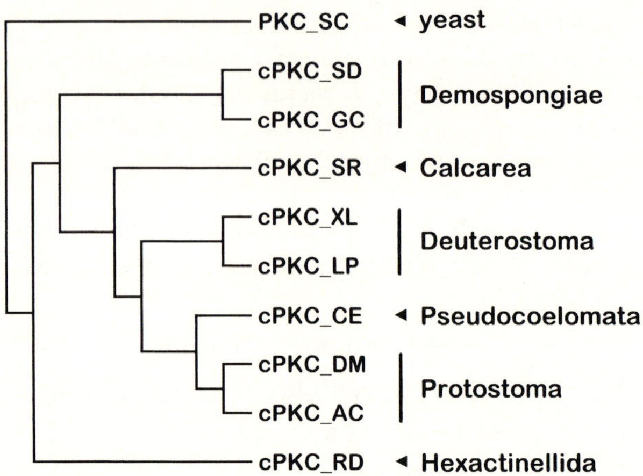

Fig. 5. Computing of the same sequences by using the algorithm described by Fitch and Margoliash (1967)

S. cerevisiae. In this multifurcational tree, the hexactinellid sponge *R. dawsoni* is separated from both the calcareous sponge *S. raphanus* and the demosponges *G. cydonium* and *S. domuncula* with high statistical significance, and the evolutionary distance between *R. dawsoni* and the demosponges is greater than that between *R. dawsoni* and the Calcarea. Furthermore, later in the phylogram the pseudocoelomate [Nematoda] *C. elegans* branches off together with the two protostomes, *A. californica* and *D. melanogaster*, as well as with the two deuterostomes, *L. pictus* and *X. laevis*. The robustness of this alignment is seen in the fact that three programs used for the construction of the tree – CLUSTAL-W program (Fig. 3), PROTPARS Protein-Parsinomy program (Fig. 4) and the algorithm of Fitch and Margoliash (Fig. 5) – resulted in the same relationships.

5.4
Phylogenetic Position of *R. dawsoni*: Analysis of the Regulatory Region

Sequence alignment using the regulatory region, including the pseudo-substrate site, the two zinc fingers, and the C2 domain present in the metazoan cPKC sequences is shown in Fig. 6A. The degree of aa identity/similarity did not vary much among the sequences: *G. cydonium* (38, 57%); *S. domuncula* (38, 62%); *S. raphanus* (45, 65%); *L. pictus* (43, 63%); and *D. melanogaster* (43, 64%).

The phylogenetic tree constructed from this alignment reflects with high statistical support the same relationships as those drawn from analysis of the catalytic domain (Fig. 6B). The hexactinellid *R. dawsoni* is at the base of the unrooted tree from which the demosponges *G. cydonium* and *S. domuncula* branch off first, and the calcareous sponge *S. raphanus* later. Again, the two

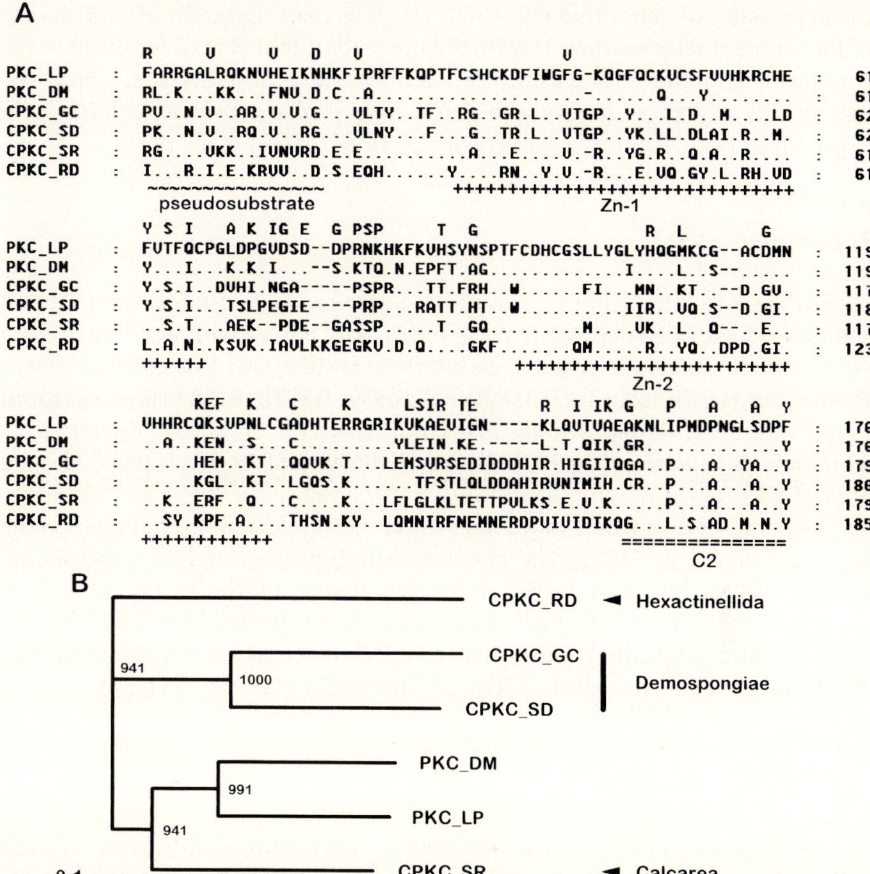

Fig. 6. A Alignment of the regulatory regions of the cPKC sequence, including the pseudosubstrate site, the two zinc fingers and the C2 domain from the metazoan species (1) Porifera; Demospongiae – *G. cydonium* (*CPKC_GC*) and *S. domuncula*, Calcarea – *S. raphanus* and Hexactinellida – *R. dawsoni*; and (2) the deuterostome (Echinodermata) *L. pictus* and the protostome (Arthropoda) *D. melanogaster*. Further description is given in the legend to Fig. 2. **B** The unrooted phylogenetic tree was constructed by neighbor-joining, and the statistical analysis was performed by bootstrapping

higher invertebrates *D. melanogaster* and *L. pictus* were found to be more closely related to the calcareous sponge.

6
The Heat Shock Proteins: Hsp70s

In a separate series of experiments, the cDNAs encoding a housekeeping protein, the 70-kDa heat shock protein (Hsp70), were isolated and characterised from Hexactinellida, Demospongiae, and Calcarea. It is well established that the expression of genes that encode the Hsp70s rises markedly after a change of the ambient temperature. Hsp70, the most abundant Hsp, is localised in the cytoplasm and the endoplasmic reticulum. These proteins share approximately 50% sequence identity with Hsp70 from prokaryotes and with Hsp70 in organelles of prokaryotic origin (reviewed in Boorstein et al. 1994).

6.1
The Sequences

cDNAs were isolated from *G. cydonium*, *S. raphanus*, and *R. dawsoni*. The nt sequences have been deposited in the EMBL data bank as follows: Hsp70 from *G. cydonium GCHSP70* [X94985; deduced aa: HSP70_GC] (Koziol et al. 1996), *R. dawsoni* Hsp70 form I *RDHSP70-1* [Y10530; HSP70_RD-1] (manuscript in prep.) and form II Hsp70 *RDHSP70-2* [Y13926; HSP70_RD-2] (Koziol et al., 1997) and *S. raphanus* Hsp70 *SRHSP70* [Y10530; HSP70_SR] Y10530 (Koziol et al., 1997).

The open reading frame of *RDHSP70-1* encodes a 655-aa-long sequence; the M_r is calculated as 72095. The characteristic bipartite nuclear targeting sequence (aa 243–259), and the Hsp70 family signatures 1 (6–13) and 2 (196–203) are present (Fig. 7). All the deduced aa sequences reported herein do have the characteristic C-terminal consensus aa EEVD or related aa which are indicative of the cytoplasmic form of Hsp70s (Boorstein et al. 1994; Fig. 7).

6.2
Phylogenetic Analysis

For phylogenetic analysis, Hsp70 sequences were truncated because of their different chain lengths. The Hsp70 sequences of *S. raphanus* (selected aa sequence 3–508) and of *R. dawsoni* (aa 3–508) were found to be highly similar to the previously reported sequence from *G. cydonium* (Koziol et al. 1996; aa 7–513) as well as to other metazoan Hsp70s, e.g. human (accession no. P17066; aa 8–513), and plant Hsp70s, e.g. the soybean *Glycine max* (P26413; aa 8–516). The percentage of identical (and similar) aa to the human sequence was as follows: *S. raphanus* 84% (93%), *R. dawsoni* 78% (89%), *G. cydonium* 78% (90%), and for the soybean *G. max*, 77% (87%). Among the sponge sequences alone, the percentage of identical (and similar) aa ranged from 80 (90%) to 82 (91%).

A high level of identity with the human sequence was also found with the corresponding truncated nt sequences: *S. raphanus* 75%, *R. dawsoni* 69%, *G. cydonium* 73%, and *G. max* 70%. Among the sponge sequences, the corresponding values were between 71 and 73%.

For the construction of a phylogenetic tree, the three sponge Hsp70 sequences were compared with the Hsp70 sequence from the bacterium *E. coli* (DNAK_ECOLI). At the aa level the bacterial Hsp70 (accession no. P04475) showed 50 to 52% identity and 67 to 69% similarity to the sponge sequences; at

```
                     T I               I                              N T
HSP70_GC    : MAKKAPAVGIDLGTTYSCVGVFQHGKVEIIANDQGNRTTPSYVAFTDSERLIGDAAKNQVA
HSP70_SR    : .----S......................................NT..............
HSP70_RD-1: .----T.I...........I...........N............N.T........S..S
HSP70_RD-2: .----T.I...........I..N.........N............N.T............
              ↑            +++++++++
                            Hsp-1
                    L    FD    Q   M   P            KIE       T T YP
HSP70_GC    : MNPTNTVFDAKRFIGRRSNDPVVSSDKKHWSFEVIDEAGRPRVRVEYKGEKKSFFAEEISS
HSP70_SR    : ...........L....FD.QT.TN.R...P.N.VSDGSK.KIE......T.T.YP.....
HSP70_RD-1: ..HA........L...KFD...IQA.M.Y.P.K..N....KIE......S.T.YP.....
HSP70_RD-2: ...S........L...FD.S..Q..M.F.P.T.VE.S...KI....A.L.T.YT.....


              L             S           A               V
HSP70_GC    : MVLTKMKETAEAYLGKTITDAVVTVPAYFNDSQRQATKDAGIISGLNILRIINEPTAAAIA
HSP70_SR    : ...V................RDVNN..I...................L.C.M.V.......
HSP70_RD-1: ...L.......D....KVS..A.........A.................VI.........
HSP70_RD-2: ...L..........S...........A..........V.....VM.........
                     ***********
                     ATP/GTP motif
             T  AG R V                  A  DD                          V
HSP70_GC    : YGLDKKHDSSEQNILIFDLGGGTFDVSILTIEEGIFEVKSTAGDTHLGGEDFDNRMUNHFI
HSP70_SR    : ......G-VG.R.V..............V.A.DD...G................L....T
HSP70_RD-1: ......V-AG.SSU.............A..D.V...R.............K..V
HSP70_RD-2: ......T-GG.RHV..............V.A.D.................V

              +++++++++   Hsp-2
                H            L                SAQ N          U F  S
HSP70_GC    : SEFKRKFKKDMSGNKRAVRRLRTACERAKRTLSSITEASIEIDSLFEGIDYYTKITRARFE
HSP70_SR    : N.....H....T....................S.Q.N........V.F..S.....
HSP70_RD-1: E.....H...I.SS...L.........T......NSAQ.N.........V.F.SS.S.....
HSP70_RD-2: G.....H...LTT....L...............HSAQ.N.......D.T.F..A.......
             ***********************      *****************************  ********
                    NTS-1       ****************************   NTS-2       ehsp70
              Q                    A LG  E              K              E
HSP70_GC    : ELCGDLFRGTLEPVEKALRDSKFDKGQIHEIVLVGGSTRIPRIQKLLQDFFNGKTLNKSIN
HSP70_SR    : ...S..............M..ANVG..E..DL..........KU............E.....
HSP70_RD-1: ..NQ....A..D.......A.LS..D.Q..........KUK.......H..E......
HSP70_RD-2: ..NQ...........S...A.LG..E..................K............E..R...
             **

                     V    A   K       V        I   A             N      T QQ
HSP70_GC    : PDEAVAYGAAIQADILTGDTSEEVQDLLLLDVTPLSLGIETAGGVMTALIKRNSTIPKKET
HSP70_SR    : ..........V..A..S..Q..A..........A.................S..N....V.T.HS
HSP70_RD-1: ..........V..A..KS...VLR.V..I..L...I.........N..D..TR..TEQQ
HSP70_RD-2: ......F...V..A..K....V...V..I..A.................N.....T...T.QQ


              Q    A    A                      D        G    E      S
HSP70_GC    : ETFTTYSDNQPGVLIQVYEGERAMTKDNNLLGKFELTGIPPAPREVPQIQVTFDIDRNGIL
HSP70_SR    : Q.....A........T................S.......G....E.....A....
HSP70_RD-1: K.....A....A.T.........L.........R.D.........G....E..LEVNSD..M
HSP70_RD-2: N.........A.T.........A.........D.........G....D....V.S....


              V             K        E D K   D   KF NE      D I    S
HSP70_GC    : NVSAADKSTGKENKITITNDKGRLSAEEIDRMUREAEQYKAADDAQRERVSAKNQLESYAF
HSP70_SR    : ....V...................K.D.E...N...R..KE.E..KD.IT...S......
HSP70_RD-1: ....VE.G.NTKKN...K.TRD.F.E..UK...D...TF.NE.N.H...IK...N....T.
HSP70_RD-2: ...GV....N.ME....K...S...ESD.K...D...KF.HE..Q..D.IG..G..T...
                                             ↑
              A  AV D       IS   KK  LD      N   S               V KU    MM
HSP70_GC    : QMKSTFEEDKVKEKVPEEDREKVISKCKEVIDWLDKNQSAEKEEFEHQQKELEGICTPIVT
HSP70_SR    : N...AV.DE.MAG.IE.S.KKTILD..N.I.A...S...............A.KU....M.
HSP70_RD-1: .A.NAI..S.AR..LS...VR..EETU.NALN...S..T.S.....Y.LG.V.KU.K..IM
HSP70_RD-2: .A...V.D....D.IT...KTN.LNIV..TVT...S..T.T.....AK..V.KU.Q.VMM


              S  G E          AG MPA  M  E  MPG M   SA     D
HSP70_GC    : KLYQAGGAPGGGMPGGMPGGMPGGMPGGFPGGAGPTSGGSTGGGSGPTIEEVD       663
HSP70_SR    : ....GA.G--A......QVEC.VVQAVQVALEP.QP.RK.IKSSD..I.S...       656
HSP70_RD-1: ...S.DT.-KLYQS.DA..AG.MPA..M..--.MPG.MPN.NAQATSVD...        655
HSP70_RD-2: ....T.AGGE.........MPG.MPGMPGGMPG.MPG.M.G.MPTNTN.DD..       658
                                                              ====
```

Fig. 7. Alignment of the deduced aa sequences of the Hsp70s from *G. cydonium* (*HSP70_GC*), *S. raphanus* (*HSP70_SR*) and *R. dawsoni* form 1 and 2 (*HSP70_RD-1 and -2*). The consensus aa are shown *above*; the identical aa with respect to the *G. cydonium* sequence are marked (·), as are gaps (−). The sites characteristic for metazoan Hsp70s, the heat shock Hsp70 proteins family signature 1 and 2 (*Hsp-1 and -2*; +++++), the ATP/GTP-binding site motif, nuclear targeting sequences (*NTS-1 and -2*), the eukaryotic Hsp70 (*ehsp70*) signature (*******) as well the Hsp70 cytoplasmic signature (====) are marked. The *arrows* delimit the part of the sequences used for the tree construction in Fig. 8

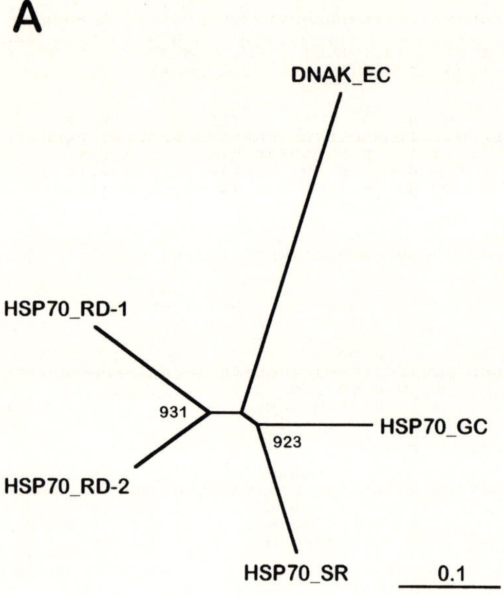

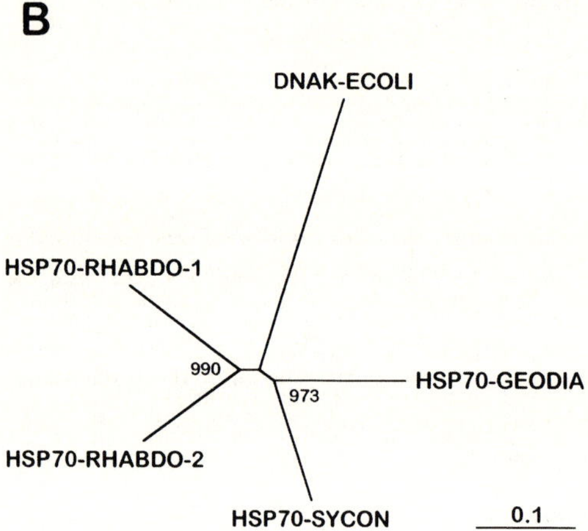

Fig. 8A,B. Phylogenetic trees using Hsp70 from the sponge sequences and the bacterial DNAK molecule inferred by neighbor-joining. The *numbers at the nodes* refer to the level of confidence as determined by bootstrap analysis (1000 bootstrap replicates). The *scale bar* indicates and evolutionary distance of 0.1 aa (in **A**) or 0.1 nt (in **B**) substitutions per position in the sequence. **A** Comparison of the truncated aa sequences from the sponges *G. cydonium* (*HSP70_GC*), *S. raphanus* (*HSP70_SR*), and *R. dawsoni* form 1 and -2 (*HSP70_RD-1 and -2*), and the bacterium *E. coli* (*DNAK_EC*; accession no. P04475). **B** Tree calculated by comparing the corresponding truncated sponge nt sequences from *G. cydonium* (*HSP70-Geodia*), *S. raphanus* (*Hsp70-Sycon*), and *R. dawsoni* (*HSP70-Rhabdo-1 and -2*) as well as from *E. coli* (*DNAK-ECOLI*)

the nt level the *E. coli* sequence (D10765) showed 50 to 52% identity with the sponge sequences. The unrooted trees, from comparisons of both aa (Fig. 8A) and nt (Fig. 8B) sequences, reflect that the sponge Hsp70 molecules are closely related. The branches with the sponge sequences show that the hexactinellid sponge is somewhat closer to *E. coli*; however, this suggestion is not statistically significant and must be further proven by analysis of nt and aa sequences of proteins which display a lower degree of identity.

7
Conclusion

Previous molecular phylogenies using 18S rRNA sequence data have concluded that the Hexactinellida share a common ancestor with the Demospongiae, and that all sponges, including the Calcarea, share a common ancestor with choanoflagellates (West and Powers 1993; Cavalier-Smith et al. 1996).

We now show that analysis of the deduced aa sequence of a cPKC from the hexactinellid sponge *R. dawsoni* and its alignment with cPKCs from the demosponges *G. cydonium* and *S. domuncula* and the calcareous sponge *S. raphanus*, provides strong statistical support that the Hexactinellida in fact evolved from a common ancestor earlier than other sponges.

Since PKC has been shown to be a phylogenetically old molecule (Kruse et al. 1996), and since the two domains of the PKC gene analysed here are not restricted to this subfamily but occur in a homologous manner in other unrelated molecules (e.g. the Cys/His effector binding domain – zinc finger – (Stabel and Parker 1991) or the catalytic domain (Hanks et al. 1988), PKCs are appropriate molecules for the phylogenetic analyses performed in this paper. Furthermore, because analysis of both parts of the cPKC sequences, (1) the catalytic domain and (2) the regulatory part of the enzyme, composed of the pseudosubstrate site, the Cys/His effector binding domains and the Ca^{2+} specific domain, resulted in the same findings, the conclusion that the Hexactinellida branched off first from a common ancestor of the phylum Porifera is strong.

This conclusion supports the proposed separation of the Porifera into two subphyla (Reiswig and Mackie 1983) which was primarily based on the difference between syncytial and cellular tissues. Since the Hexactinellida are syncytial animals (Mackie and Singla 1983) and since the present analysis indicates that they branched off from a common ancestor earlier than other sponges, this could imply that multicellularity came about by division of a multinucleate syncytium rather than by aggregation of formerly single cells. However, it is also possible that the syncytial nature of the Hexactinellida reflects a reduction of a previously multicellular stage by fusion to form a syncytium. The existing data, that aggregates from hexactinellids form syncytia by fusion of uni- and multinucleate cells (Pavans de Cecatty and Mackie 1982; Leys 1995) supports the latter view. To clarify this question it will be necessary to examine different hexactinellid sponges with respect to adhesion molecules and receptors, and, most importantly, to re-examine the embryology of the Hexactinellida of which very little is known at present (see Boury-Esnault and Vacelet 1994).

Additional support for an early origin of the Hexactinellida comes from paleontological evidence, which shows that the Hexactinellida were present in the early Cambrian while the Calcarea and the Demospongiae arose later towards the Middle Cambrian (Reitner and Mehl 1995). The oldest sponge spicules found to date (approximately 600 million years old) come from China, and are mainly monaxial spicules. Some, however, also have definite "crosses", which are typical of triaxones from hexactinellids (Steiner et al. 1993).

A further outcome of our study is the finding that the cPKC sequence from the calcareous sponge *S. raphanus* is related more closely to the cPKC sequence from the protostome *L. pictus* and to that from the deuterostome *D. melanogaster* than it is to the cPKC sequence from the two demosponges. Although analysis of earlier rRNA sequence data revealed a less clear relationship (Lafay et al. 1992), our findings are in agreement with recent 18S rRNA sequence data (Cavalier-Smith et al. 1996), and are supported by strong bootstrap values for phylogenetic trees formed by analysis of both regions of the PKC gene.

This analysis has further shown that comparison of genes coding for proteins which are highly conserved cannot resolve the relationships between the classes of the Porifera, the earliest metazoan phylum. For this purpose it will be necessary to examine proteins such as protein kinase C, which have been demonstrated to have a higher rate of evolutionary change.

In conclusion, within the phylum Porifera, the class Hexactinellida diverged first from a common ancestor, while Calcarea and Demospongiae appeared later. Furthermore, the cPKC sequences from higher invertebrates are more closely related to the cPKC from Calcarea.

Acknowledgements. The studies performed in the authors' laboratory were supported by grants from the Deutsche Forschungsgemeinschaft [Mü 348/12-1] and the International Human Frontier Science Program [W.E.G. Müller; RG-333/96-M].

References

Anderson PAV (1980) Epithelial conduction: its properties and functions. Prog Neurobiol 15:161–203

Böger H (1988) Versuch über das phylogenetische System der Porifera. Meyniana 40:143–154

Boorstein WR, Ziegelhoffer T, Craig EA (1994) Molecular evolution of the HSP70 multigene family. J Mol Evol 38:1–17

Boury-Esnault N, De Vos L (1988) *Caulophacus cyanae*, n. sp., une éponge hexactinellide des sources hydrothermales. Biogéographie du genre *Caulophacus* Schulze, 1887. Oceanol Acta 8:51–60

Boury-Esnault N, Vacelet J (1994) Preliminary studies on the organisation and development of a hexactinellid sponge from a Mediterranean cave, *Oopsacas minuta*. In: van Soest RWM, van

Kempen TMG, Braekman JC (eds) Sponges in time and space. Balkema Press, Rotterdam, pp 407–415

Cavalier-Smith T, Allsopp MTEP, Chao EE, Boury-Esnault N, Vacelet J (1996) Sponge phylogeny, animal monophyly, and the origin of the nervous system: 18S rRNA evidence. Can J Zool 74:2031–2045

Christen R, Ratto A, Baroin A, Perasso R, Grell KG, Adoutte A (1991) An analysis of the origin of metazoans, using comparisons of partial sequences of the 28S RNA, reveals an early emergence of triploblasts. EMBO J 10:499–503

Dayhoff MO, Schwartz RM, Orcutt BC (1978) A model of evolutionary change in protein. In: Dayhoff MO (ed) Atlas of protein sequence and structure, vol 5, suppl 3. National Biomedical Research Foundation, Washington, DC, pp 345–352

Felsenstein J (1993) PHYLIP, version 3.5. University of Washington, Seattle

Field KG, Olsen GJ, Lane DJ, Giovannoni SJ, Ghiselin MT, Raff EC, Pace NR, Raff RA (1988) Molecular phylogeny of the animal kingdom. Science 239:748–753

Fitch WM, Margoliash E (1967) Construction of phylogenetic trees. Science 155:279–284

Gamulin V, Rinkevich B, Schäcke H, Kruse M, Müller IM, Müller WEG (1994) Cell adhesion receptors and nuclear receptors are highly conserved from the lowest Metazoa (marine sponges) to vertebrates. Biol Chem Hoppe Seyler 375:583–588

Hadzi J (1963) The evolution of the Metazoa. Pergamon Press, Oxford

Haeckel E (1868) Natürliche Schöpfungsgeschichte. Reimer, Berlin

Hanks SK, Quinn AM, Hunter T (1988) The protein kinase family: conserved features and deduced phylogeny of the catalytic domains. Science 241:42–52

Hanson ED (1977) Origin and early evolution of animals. Wesleyan University Press, Middletown, 1–670

Hardie G, Hanks S (1995) The protein kinase fact book: protein-tyrosine kinases. Academic Press, London

Higgins DG, Sharp PM (1988) CLUSTAL: a package for performing multiple sequence alignment on a microcomputer. Gene 73:237–244

Hunter T (1991) Protein kinase classification. Methods Enzymol 200:3–37

Hyman LH (1940) The invertebrates: Protozoa through Ctenophora, vol 1. McGraw-Hill, New York, pp 248–696

Inglis WG (1985) Evolutionary waves: patterns in the origins of animal phyla. Aust J Zool 33:153–178

Koziol C, Wagner-Hülsmann C, Cetkovic H, Gamulin V, Kruse M, Pancer Z, Schäcke H, Müller WEG (1996) Cloning of the heat-inducible biomarker, the cDNA encoding the 70-kDa heat shock protein, from the marine sponge *Geodia cydonium*: response to natural stressors. Mar Ecol Prog Ser 136:153–161

Koziol C, Leys SP, Müller IM, Müller WEG (1997) Cloning of Hsp70 genes from the marine sponges *Sycon raphanus* (Calcarea) and *Rhabdocalyptus dawsoni* (Hexactinellida). An approach to solve the phylogeny of sponges. Biol J Linnean Soc 62:581–592

Kruse M, Gamulin V, Cetkovic H, Pancer Z, Müller IM, Müller WEG (1996) Molecular evolution of the metazoan protein kinase C multigene family. J Mol Evol 43:374–383

Lafay B, Boury-Esnault N, Vacelet J, Christen R (1992) An analysis of partial 28S ribosomal RNA sequences suggests early radiation of sponges. BioSystems 28:139–151

Lake JA (1990) Origin of Metazoa. Proc Natl Acad Sci USA 87:763–766

Leys SP (1995) Cytoskeletal architecture and organelle transport in giant syncytia formed by fusion of hexactinellid sponge tissues. Biol Bull 188:241–254

Leys SP (1996) Cytoskeletal architecture, organelle transport, and impulse conduction in hexactinellid sponge syncytia. PhD Thesis, University of Victoria

Leys SP (1997) Sponge cell culture: a comparative evaluation of adhesion to a native tissue extract and other culture substrates. Tissue Cell 29:77–87

Leys SP, Mackie GO (1997) Electrical recording from a glass sponge. Nature 387:29–30

Mackie Go (1965) Conduction in the nerve-free epithelia of siphonophores. Am Zool 5:439–453

Mackie GO (1981) Plugged syncytial interconnections in hexactinellid sponges. J Cell Biol 91:103a

Mackie Go, Singla CL (1983) Studies on hexactinellid sponges. I. Histology of *Rhabdocalyptus dawsoni* (Lambe, 1873). Philos Trans R Soc Lond [Biol] 301:365–400

Mehl D, Reiswig HM (1991) The presence of flagellar vanes in choanomeres of Porifera and their possible phylogenetic implications. Z Zool Syst Evol Forsch 29:312–319

Morris PJ (1993) The developmental role of the extracellular matrix suggests a monophyletic origin of the kingom Animalia. Evolution 47:152–165

Müller WEG (1995) Molecular phylogeny of Metazoa (animals): monophyletic origin. Naturwissenschaften 82:321–329

Müller WEG (1997a) Molecular phylogeny of Eumetazoa: experimental evidence for monophyly of animals based on genes in sponges (Porifera). Prog Mol Subcell Biol 19:89–132

Müller WEG (1997b) Origin of metazoan adhesion molecules and adhesion receptors as deduced from their cDNA analyses from the marine sponge *Geodia cydonium*. Cell Tissue Res 289:383–395

Müller WEG, Müller IM, Gamulin V (1994) On the monophyletic evolution of the Metazoa. Braz J Med Biol Res 27:2083–2096

Nishizuka Y (1992) Intracellular signalling by hydrolysis of phospholipids and activation of protein kinase C. Science 258:607–614

Pavans de Cecatty M, Mackie GO (1982) Génèse et évolution des interconnections syncytiales et cellulaires chez une éponge Hexactinellide en cours de réaggregation après dissociation in vitro. C R hebd Séances Acad Sci 294:939–944

Reiswig HM, Mackie GO (1983) Studies on hexactinellid sponges. III. The taxonomic status of the Hexactinellida within the Porifera. Philos Trans R Soc Lond [Biol] 301:419–428

Reitner J, Mehl D (1995) Early paleozoic diversification of sponges: new data and evidence. Geol Paläontol Mitt Innsbruck 20:335–347

Rodrigo AG, Bergquist PR, Bergquist PL, Reeves RA (1994) Are sponges animals? An investigation into the vagaries of phylogenetic interference. In: von Soest RWM, von Kempen TMG, Braekman JC (eds) Sponges in time and space. Balkema Press, Rotterdam, pp 47–54

Stabel S, Parker PJ (1991) Protein kinase C. Pharmacol Ther 51:71–95

Steiner M, Mehl D, Reitner J, Erdtmann BD (1993) Oldest entirely preserved sponges and other fossils from the Lowermost Cambrian and a new facies reconstruction of the Yangtze Platform (China). Berl Geowiss Abh [E] 9:293–329

Wachtmann D, Stockem W, Weissenfels N (1990) Cytoskeletal organization and cell organelle transport in basal epithelial cells of the freshwater sponge *Spongilla lacustris*. Cell Tissue Res 261:145–154

West L, Powers D (1993) Molecular phylogenetic position of hexactinellid sponge in relation to the Protista and Demospongiae. Mol Mar Biol Biotechnol 2:71–75

Willmer P (1994) Invertebrate relationships: pattern in animal evolution. Cambridge University Press, Cambridge

Structure and Evolution of Genes Encoding Polyubiquitin in Marine Sponges

V. Gamulin and L. Lukic[1]

1
Introduction

Molecular biology, with its rapid advances in new techniques, has drastically influenced evolutionary investigations and a new field, molecular evolution, was recently established. In the previous period traditional evolutionary biologists based their investigations upon comparative anatomy, embryology and paleontology. The recognition of the enormous phylogenetic information in nucleotide (nt) and amino acid (aa) sequences directed the efforts of molecular evolutionists toward the extraction and analysis of that information. Comparative analysis of conserved genes (proteins) in different contemporary organisms provides information about ancient events in nature and greatly facilitates reconstruction of the biological history of the living world. The nature of the first common ancestor of all living organisms is one of the central, still unsolved problems in biology.

In this article, results obtained from analyses of polyubiquitin genes from marine sponges are reviewed and discussed in connection with the phylogenetic placement of sponges (Porifera).

2
Ubiquitin

2.1
Role in Protein Degradation

Ubiquitin is a small, abundant and stable globular protein, composed of 76 aa, with Mr of around 8.5 kDa. It plays a major role in the normal breakdown of unwanted proteins in the cell (reviewed by Mayer et al. 1991; Finley and Chau 1991; Doherty and Mayer 1992; Hershko and Ciechanover 1992; Jentsch 1992). Ubiquitin is also involved in a variety of other fundamental cellular processes such as DNA repair (Jentsch et al. 1987), maintenance of chromatin structure

[1]Department of Molecular Genetics, Rudjer Boskovic Institute, Bijenicka 54, 10000 Zagreb, Croatia

Progress in Molecular and Subcellular Biology, Vol. 21
W.E.G. Müller (Ed.)
© Springer-Verlag Berlin Heidelberg 1998

(Finley and Varshavsky 1985), cell cycle control (Goebl et al. 1988; Glotzer et al. 1991), regulation of gene expression (Nickel et al. 1989), cell response to stress (Bond and Schlesinger 1986) and ribosome biogenesis (Finley et al. 1989).

Investigation of the crystal structure indicated that ubiquitin is a tightly packed globular protein with a protruding C-terminus (Vijay-Kumar et al. 1987a) of four residues forming a "tail". Extensive hydrogen bonding exists within the ubiquitin molecule, suggesting a rigid structure. This is most probably the reason for its remarkable thermal stability and its resistance to a broad pH range.

In non-lysosomal- and ATP-dependent proteolysis mediated by this small protein, a complex process of selection and marking of proteins for degradation takes place. Catalysed by a family of enzymes, ubiquitin is covalently attached to target proteins, and ubiquitinated proteins are subsequently rapidly degraded by a large molecular mass protease. Ubiquitin is released during protein degradation to be used in further rounds of protein catabolism. The conserved C-terminus of ubiquitin participates in ubiquitin protein conjugation. An isopeptide bond is formed between the C-terminal glycine residue of ubiquitin and the ε-amino group of available lysine residues in the target protein. Target proteins may be ubiquitinated multiple times and a chain of up to 20 ubiquitins can be attached to a single lysine residue of a target protein. Isopeptide bonds between ubiquitin units are formed involving lysine 48 and C-terminal glycine.

Ubiquitin plays an important role in protein catabolism both in the nucleus and in the cytoplasm. Stable conjugates have been identified with proteins from the nucleus (i.e. histones, cyclins, protooncogenes) and the cell surface (cell adhesion proteins). Although ubiquitin has been extensively studied as a cofactor for non-lysosomal protein degradation, some results have suggested a role for ubiquitin in lysosomal proteolysis (Laszlo et al. 1990), particularly during nutrient deprivation and heat stress (Gropper et al. 1991). It would appear that ubiquitin mediated protein degradation plays a key role in eukaryotes, especially in the breakdown of short-living, needless and abnormal proteins.

The N-terminus residue of certain proteins may determine whether that protein is a good acceptor for ubiquitin-mediated rapid degradation (Bachmair et al. 1986). Rapidly degraded proteins have basic aa's at the N-terminus (Arg, Lys and His). However, the majority of proteins may be recognized by the ubiquitination system by signals other than those at N-end.

After heat-shock of organisms and the production of heat-damaged proteins, during the "heat-shock response" a wave of ubiquitination is observed and heat-damaged proteins are targeted for degradation by the ubiquitin pathway. Ubiquitin itself is a heat-shock protein and elevated temperatures lead to increased expression of polyubiquitin genes.

Enzymes involved in the ubiquitin-protein conjugation are ubiquitin-activating enzyme (E1) and a family of ubiquitin-conjugating enzymes (E2). In some cases an additional enzyme (E3) is required for the ligation of ubiquitin to target proteins. Proteasome, or large molecular weight ($>$1000 kDa) multicatalytic protease complex, (also called ubiquitin conjugate degrading enzyme) is involved in the recognition of multi-ubiquitinated protein substrates and starts the degradation of ubiquitinated proteins (Eytan et al. 1989).

2.2
Occurrence in Nature

Ubiquitin was originally named for its presence in all cell types (Goldstein et al. 1975), although later studies failed to find it in bacteria. It is, however, found in all eukaryotic cells and ubiquitin-mediated proteolysis has been demonstrated in protozoan organisms, yeasts and plants, as well as mammalian cells and tissues (e.g. muscle, liver, heart, eye lens). Recent discovery of a ubiquitin-mediated proteolytic pathway in the archaebacterium *Thermoplasma acidophilum* indicates that this function is ancestral to all eukaryotes (Wenzel and Baumeister 1993; Wolf et al. 1993).

2.3
Conservation of the Primary Structure

Despite the widespread distribution of ubiquitin in nature and its numerous important biological functions, this small and abundant cellular protein was indentified and characterized only 20 years ago (Goldstein et al. 1975). Early investigators noticed its remarkable structure conservation among different organisms (Schlesinger and Goldstein 1975). During the last 20 years, the primary structure of ubiquitin was elucidated in dozens of eukaryotic organisms. Ubiquitin is now thought to be the most highly conserved protein of all known eukaryotic proteins (Sharp and Li 1987a). The C-terminal end in all analysed ubiquitins always has the same conserved amino acids sequence: Val_{70}-Leu-Arg-Leu-Arg-Gly-Gly_{76}. Only in the sponge, *Geodia cydonium*, has one of the repeats of the polyubiquitin gene changed Leu to Val by mutation in codon 71. The majority of the other aa's are also fully conserved. X-ray crystallographic structures of human, oat and yeast ubiquitins are nearly identical (Vijay-Kumar et al. 1987b). The remarkable conservation of ubiquitin aa sequences is also evident from comparisons of coding regions from the fungal, plant and animal kingdoms. The aa sequence of animal ubiquitin is identical from insects (*Drosophila*) to humans (Jentsch et al. 1991). Similarly, all functional ubiquitin genes from higher plants that have been analysed encode identical proteins, which differ from animal ubiquitin in only three positions, from yeast ubiquitin in two positions and from Chlamidomonas ubiquitin in one position (Callis et al. 1995). Slightly higher differences were found in different protist groups. However, 62 out of 76 aa in ubiquitin are identical throughout the phylogenetic tree (Wöstmann et al. 1992; Wray and DeSalle 1994), and the majority of substitutions are located at only 5–6 positions in the protein sequence (Table 1).

Table 1. Variations of amino acids in ubiquitin. Amino acid sequences of ubiquitin from various species are compared with the "animal" ubiquitin which includes mammals, birds, amphibia, fish and insects. Only those residues which differ from the "animal" protein are shown

Residue position

Organisms	16	19	22	24	28	57
Animals	Glu	Pro	Thr	Glu	Ala	Ser
Euplotes	Asp	Gln	Thr	Asp	Thr	Ala
Dyctiostelium	Glu	Gly	Asn	Glu	Thr/Ala	Ala
Saccharomyces	Glu	Ser	Thr	Asp	Ser	Ser
Caenorhabditis	Glu	Ala	Thr	Glu	Ala	Ser
Arabidopsis	Glu	Ser	Thr	GLu	Ala	Ala
Geodia	Glu	Ala	Thr	Asp	Ala	Ser
Suberites	Glu	Ala	Thr	Glu	Ala	Ser
Sycon	Glu	Pro	Thr	Asp	Ala	Ser

3
Ubiquitin Genes

The genes encoding ubiquitin were not elucidated until 1984 (Özkaynak et al. 1984; Dworkin-Rastl et al. 1984) and then produced several surprises. All ubiquitin genes discovered so far encode ubiquitin precursors. Three classes of ubiquitin genes can be found in all eukaryotes and the overall structures and sequences of all three classes are well conserved in nature. The simplest version of ubiquitin genes encodes ubiquitin plus one additional C-terminal aa which blocks C-terminal glycine (Jentsch et al. 1991). However, the prevailing majority of ubiquitin genes encode fusion proteins and the initial translation products are processed into ubiquitin monomers by specific C-terminal hydrolases (Özkaynak et al. 1987; Mayer and Wilkinson 1989). This coding arrangement is unique so far among eukaryotic genes. Ubiquitin precursors are often encoded by multicopy gene families and multiple ubiquitin genes are characteristic of most eukaryotes investigated to date.

3.1
Class I and II: Ubiquitin Fusion Genes

Class I and II genes (Jentsch et al. 1991) code of N-terminal ubiquitin fused with ribosomal proteins of a large subunit (52 aa) or a small subunit (76–81 aa). These ubiquitin fusion genes are highly conserved among species and have been extensively studied in yeast (Finley et al. 1989). Hybrid proteins are cleaved by specific proteases (Baker et al. 1992) to yield free ubiquitin and ribosomal proteins. The transient association between ubiquitin and ribosomal proteins promotes their incorporation into nascent ribosomes and is required for efficient ribosome biogenesis. Ubiquitin acts here as a "chaperone" which, in covalent association with ribosomal proteins, promotes the formation of specific cellular structures (Finley et al. 1989).

3.2
Class III: Polyubiquitin Genes

Class III ubiquitin genes (Jentsch et al. 1991) code for polyubiquitin proteins where several ubiquitin coding regions are lined up head to tail (Özkaynak et al. 1984; Dworkin-Rastl et al. 1984). The number of ubiquitin units in these precursors differs from organism to organism and from locus to locus. One clear exception found to date is *Giardia lamblia* with a single ubiquitin gene (Krebber et al. 1994). On the other side there is *Trypanosoma cruzi* with up to 50 ubiquitin repeats in the polyubiquitin gene (Swindle et al. 1988).

In most organisms only some of the ubiquitin polygenes belonging to the multicopy gene family have been studied in detail. In the primitive and well studied eukaryotic organism, yeast, polyubiquitin is encoded by a single gene, UBI4, and the fusion protein consists of five ubiquitin units, followed by an additional aa at the C-terminus of the last repeat (Finley et al. 1987; Özkaynak et al. 1984). This gene is strongly expressed under stress conditions (including heat stress) and only under these conditions is UBI4 essential for cell viability (Finley et al. 1987).

In much more complex organisms, such as the flowering plant *Arabidopsis thaliana*, the total complement of the ubiquitin gene family consists of 14 members, including five polyubiquitin genes (Callis et al. 1995). Other ubiquitin genes belong to class I and class II ubiquitin fusion genes or ubiquitin-like genes. Five *A. thaliana* polyubiquitin genes contain three to six tandem repeats of the ubiquitin coding region and they all code for the same protein, identical to the ubiquitin of higher plants and differing from the animal consensus ubiquitin sequence at three positions. However, all five polyubiquitin genes differ in synonymous substitutions, number of ubiquitin coding regions, number and nature of non-ubiquitin C-terminal aa(s) and chromosome location.

Polyubiquitin genes generally do not contain introns. Among all sequenced genes, introns were found only in the *Caenorhabditis elegans* major polyubiquitin locus (Graham et al. 1989) and in one of the *A. thaliana* pseudoubiquitin genes (Callis et al. 1995).

The presence of multiple, expressed polyubiquitin genes that differ in repeat number and sequence (like in *A. thaliana*) is typical in eukaryotes. One exception, already mentioned, is *Saccharomyces cerevisiae* with only one polyubiquitin gene. It is however known that, for example, *Dictyostelium* expresses at least five different polyubiquitin genes (Ohmachi et al. 1989) and at least two expressed polyubiquitin genes (with nine and three repeats) have been identified in humans (Wiborg et al. 1985; Baker and Board 1987).

3.3
Molecular Evolution of Ubiquitin Genes

Because all analysed eukaryotes (except *Giardia*) contain at least one gene with tandem repeats within the ubiquitin coding region, this polygene structure is most probably very ancient and preceeds the speciation of eukaryotic organisms. Later events, which included gene duplications and translocations, created multiple polyubiquitin gene families. One would therefore expect that the differences between gene repeats within one polyubiquitin gene (one locus) are higher than the differences between orthologous gene repeats from different loci, or homologous gene repeats from two different species. Orthologous (homologous) ubiquitin gene repeats – created by gene duplication(s) and speciation – share a more recent common ancestor. However, it is a fact that the highest homology exists between the repeats within the locus (Sharp and Li 1987a,b) and the next best homology was observed between repeats from the different loci within the same species. In general, all ubiquitin repeat units from expressed polyubiquitin genes within one species (whether from the same or different loci) encode the same protein. Exceptions have so far been found only in *G. cydonium* (Pfeifer et al. 1993a) and two protists, *Tetrahymena pyriformis* (Neves et al. 1990) and *Trichomonas vaginalis* (Keeling and Doolittle 1995). Thus, ubiquitin repeats within species seem to evolve "in concert" and concerted evolution arose from "communication" among the members of a repeat family (Sharp and Li 1987a).

In ubiquitin genes, concerted evolution involves both unequal crossing-over (Ohta and Dover 1983) and (at a much lower rate) gene conversion, involving family members at nonhomologous chromosomal sites (Nagylaki 1984; Ohta 1984). The efficacy of within and between locus concerted evolution is very different and "communication" is much less effective between loci (Sharp and Li 1987b). A high degree of divergence exists between repeat units within ubiquitin loci, suggesting that concerted evolutionary events, although obviously present in ubiquitin genes, occur infrequently (Sharp and Li 1987b). Concerted evolution has also been found in other repeated sequence families and the best known example is the family of ribosomal RNA genes (Ohta 1980). However, because ubiquitin genes are under most stringent structural constraints, these genes provide the best model genes for the study of concerted evolution of repeats within the cluster, unequaled by any other known gene with internal repeats (Sharp and Li 1987b).

4
Ubiquitin in Marine Sponges

The answer to the question about the origin of Porifera (monophyletic or polyphyletic in connection to all other Metazoa), came only recently, after the application of molecular biological techniques, which made it possible to take nucleotide sequence data into consideration for the taxonomy of animals.

In our group we adopted an approach which relies on the identification and analysis of phylogenetically conserved genes (proteins) in the marine sponges from the subphylum Cellularia (classes Calcispongiae and Demospongiae). All sponges were collected from the Adriatic Sea (Istria, Croatia). Genes, selected for analysis, encode proteins specific for multicellular organisms and are well conserved among different groups of Metazoans (e.g. lectins, tyrosine kinases). In addition, well conserved proteins, known to be present in all Eukarya (ubiquitin, protein kinase C), or even in prokaryotic organisms (heat shock protein), were also analysed. The results were mostly obtained from the sponge *G. cydonium* (Demospongiae). All our results speak in favour of the monophyletic origin of Metazoa, i.e. common origin of all Metazoans, including sponges (Gamulin et al. 1994; Müller 1995; Müller et al. 1995a,b). Deduced aa sequences of sponge proteins display significant homology to the corresponding proteins from different Metazoa, including mammals. However, sponge proteins always branch off first from the phylogenetic tree, indicating a very ancient split of Porifera from the ancestral organism, common to all Metazoa. The approximate time of this split, estimated to be over 600 million years ago (Schäcke et al. 1994), corresponds very well with paleontological data.

In the following Sections, only results obtained from the analysis of polyubiquitin genes in marine sponges will be discussed in more detail.

4.1
Polyubiquitin Gene from *Geodia cydonium*

A cDNA library of *G. cydonium* was prepared from a specimen collected near Rovinj, northern Adriatic Sea, Croatia, (Pfeifer et al. 1993a). mRNA, extracted from unselected sponge tissue, was used for the synthesis of the cDNA and the library was prepared in a λZAP expression vector (Short et al. 1988). Human ubiquitin cDNA probe (Baker and Board 1987) was used for the identification of sponge (poly) ubiquitin cDNA clones. Five independent clones were identified and analysed and all analyses resulted in the same sequence. All clones contained a 1.6-kb-long cDNA insert, which encoded a polyubiquitin precursor protein with six highly homologous sequences for ubiquitin, joined head-to-tail, without spacers, as found in all other analysed eukaryotes (Pfeifer et al. 1993a). cDNA clones encoding ubiquitin fused to ribosomal proteins were not identified in this cDNA library. Northern hybridization analysis of sponge mRNAs, using a homologous cDNA probe, also identified only one size of the ubiquitin mRNAs (1.6kb long; Pfeifer et al. 1993a). The obvious conclusion was that in the sponge specimen used for the preparation of the cDNA library, only one polyubiquitin gene was highly expressed (only one gene exists?). Identification of potential additional ubiquitin genes by Southern hybridization was not performed. Therefore, the total number of polyubiquitin genes in *C. cydonium* can theoretically be more than one.

The nt sequence of sponge polyubiquitin cDNA (termed GCUBI1; EMBL accession number X70917) is shown in Fig. 1. The open reading frame is 1371 bp long and starts with the ATG codon for Met; this initiation codon confirms with the Kozak consensus sequence CAGCATG (Kozak 1984) for the translational start site in eukaryotic mRNA. The stop codon is TAA. The typical signal for a polyadenylation site, AATAAA (Zarkower et al. 1986), is missing in GCUBI1_cDNA. Six ubiquitin repeats in the sponge polyubiquitin gene encode the same protein (with one exception in repeat 4, see later). However, on the nucleotide level the sequences of the six repeats differ considerably. All changes (except one change in the repeat 4) are silent substitu-

```
GCUBI1                         -103 GCCAAGGAAACGCACGTCCGAGTTCACTCGAATCTTCTCGCTT
GCUBI1                  -60 TCGAAGCGCCCCGAAACTCATTTCTCAACCGTCGAAGCCTACTTCAAGAAAGCTTTCAAC
Group I:
GCUBI1_1 nt        1 ATGCAAATCTTCGTCAAGACACTCACCGGAAAGACCATCACACTAGAGGTCGAGGCCAGC
GCUBI1_3         457 .............C.....G.....C.........A........C...........T..C
GCUBI1_5         913 .............C.....G.....A.........C.........A...........T..C
Group II:
GCUBI1_2         229 .............T.....G.....C.........A........C...........T..C
GCUBI1_4         685 .............T.....G.....C.........A........C...........T..C
GCUBI1_6        1141 .............C.....G.....C.........C.........A...........T..T
GCUBI1   aa        1 M  Q  I  F  V  K  T  L  T  G  K  T  I  T  L  E  V  E  A  S
Group I:
GCUBI1_1 nt       61 GATACCATCGAGAACGTGAAGGCCAAGATCCAGGACAAGGAGGGGATCCCGCCCGACCAG
GCUBI1_3         517 ..C.....C.....C........T................A..C.............
GCUBI1_5         973 ..C.....C.....C........C................A..G.............
Group II:
GCUBI1_2         289 ..C.....C.....C........T................A..C.............
GCUBI1_4         745 ..C.....T.....C........T................A..C.............
GCUBI1_6        1201 ..C.....C.....T........C................A..G.............
GCUBI1   aa       21 D  T  I  E  N  V  K  A  K  I  Q  D  K  E  G  I  P  P  D  Q
Group I:
GCUBI1_1 nt      121 CAGCGTCTCATCTTCGCCGGCAAGCAGCTCGAAGACGGTCGCACACTGAGCGACTACAAC
GCUBI1_3         577 .............C..C..C..G.....C.....C..T..C..A..G.....C......
GCUBI1_5        1033 .............T..C..A..G.....C.....C..T..T..A..G.....C......
Group II:
GCUBI1_2         349 .............C..T..C..G.....T.....T..C..C..A..G.....C......
GCUBI1_4         805 .............C..T..C..G.....T.....T..C..C..G..G.....C......
GCUBI1_6        1261 .............T..T..C..A.....C.....C..C..C..A..A.....T......
GCUBI1   aa       41 Q  R  L  I  F  A  G  K  Q  L  E  D  G  R  T  L  S  D  Y  N
Group I:
GCUBI1_1 nt      181 ATCCAAAAGGAATCTACCCTTCACCTTGTACTCCGTCTGCGTGGTGGT
GCUBI1_3         637 ..C........ATCT.........C..T..AC....T...........T
GCUBI1_5        1093 ..C........ATCC.........T..T..GC....C...........T
Group II:
GCUBI1_2         409 ..T........GAGC.........T..C..GC....C...........T
GCUBI1_4         865 ..T........GAGC.........T..T..CG....C...........T
GCUBI1_6        1321 ..T........GAGC.........T..C..GC....T...........GTTC 1371 TAA
GCUBI1   aa       61 I  Q  K  E  S65 T  L  H  L  V  L/V R  L  R  G  G
GCUBI1         +1375 ATCCCACGTGAAAGAACTCTGGATTGATTGAACTTGGACAACAGTAGTGTAGTTCTTGCT
GCUBI1         +1435 GTCACAACATGCACACAAAGCTTGTATGCTGTATAATACCTGTTACCCTCGTAAAAAAAA
GCUBI1         +1495 AAAAAAAA
```

Fig. 1. Nucleotide sequence of *G. cydonium* polyubiquitin cDNA (GCUBI1) arranged according to the level of between-repeat homologies. The deduced aa sequence of ubiquitin is also shown; the Ser at position 65 is marked. *Dots* indicate identities throughout the six repeats

tions, which do not affect the protein structure. Detailed analysis of the silent mutations in the *G. cydonium* polyubiquitin gene will be presented in Section 4.1.2.

The amino acid sequence of the *G. cydonium* ubiquitin, deduced from the cDNA, differs from the human ubiquitin at only one position: aa at position 19 is Pro in humans (and all higher Metazoa from *Drosophila* to man; see Table 1) and in *G. cydonium* Ala was found at position 19. Ubiquitin with the same primary structure as in *G. cydonium,* was also found in the nematode *C. elegans* (Graham et al. 1989). One replacement substitution was found in six. *G. cydonium* repeats: in the GCUBI1_4 repeat (see Fig. 1) Leu at position 71, found in all analysed ubiquitins (Jentsch et al. 1991; Finley and Varshavsky 1985), is replaced by Val. The C-terminal consensus ubiquitin sequence: Val_{70}-Leu-Arg-Leu-Arg-Gly-Gly_{76} is essential for the function of ubiquitin. It is therefore very possible that the ubiquitin encoded by GCUBI1_4 repeat in *G. cydonium* is not a functional protein. As in most other polyubiquitin genes, the last repeat, GCUBI1_6, encodes one extra aa (Phe) and an additional Phe at the C-terminus of the last repeat is also encoded i.e. in the polyubiquitin gene from the pig, *Sus scrofa* (Einspanier et al. 1987).

4.1.1
Expression of the Polyubiquitin Gene in Geodia cydonium

The level of polyubiquitin mRNA is almost identical in two sponge compartments – cortex and medulla (Pfeifer et al. 1993a). However, detectable amounts of ubiquitin proteins are present only in the cortex of the sponge tissue and not in the medulla region. This finding might mean that ubiquitin mRNA is not, or only to a small extent, translated in the medulla region. To the best of our knowledge, the expression of ubiquitin in other systems is primarily regulated at the transcriptional level (Schwartz et al. 1990). In the *G. cydonium* system, translation efficiency of ubiquitin mRNA appears to be under fine control (Pfeifer et al. 1993a).

In vitro studies with dissociated sponge cells revealed that the homologous aggregation factor (AF) (Müller and Zahn 1973; Henkart et al. 1973) causes a strong increase in the steady state level of mRNA coding for ubiquitin and consequently a drastic increase in ubiquitin protein synthesis (Pfeifer et al. 1993a). Sponge AF is an adhesion molecule for cortex cells, which also functions as a mitogen (Müller et al. 1990) and modulates morphogen-induced differentiation of sponge cells (Biesalski et al. 1992). Sponge lectin (Pfeifer et al. 1993b), another mitogen and aggregation molecule specific for sponge medulla cells (Müller et al. 1988; Gramzow et al. 1989), failed to display the same effect in isolated cells. These data suggest that ubiquitin may play a role in sponge morphogenesis (Pfeifer et al. 1993a).

Ubiquitin isolated from *G. cydonium* was found to initiate protein degradation in the heterologous reticulocyte system in the same manner as bovine ubiquitin (Pfeifer et al. 1993a).

4.1.2
Phylogenetic Relationships of Ubiquitin Repeats in the Polyubiquitin Gene from Geodia cydonium

Due to the extreme evolutionary conservation of the ubiquitin aa sequence, this protein is apparently not suitable for phylogenetic studies. However, because of the different codon usage, gene sequences can diverge considerably without changes in the protein structure, and therefore polyubiquitin genes are quite suitable models for studying mutational events which occurred in the living world during ubiquitin evolution.

4.1.2.1
Enigma with Serine Codons

In *G. cydonium* polyubiquitin gene repeats GCUBI1_1 to GCUBI1_6 (228 bP each), changes in the nt sequence were found altogether at 35 positions (Müller et al. 1994; Fig. 1). However, the number of nt changes between any two repeats is much lower (maximum 22, as found between GCUBI1_1 and GCUBI1_2). Only one change in the GCUBI1_4 (C to G at the first position of codon 71) caused a transition from Leu to Val, and all other changes are silent substitutions, mostly at the third position of the affected codons. However, changes in Ser codons at position 65 look like double substitutions involving first and second bases of the respective codons (Fig. 1). A similar situation (at serine codon 20) was also found in *D. discoideum* and was explained as a simultaneous double mutation (Sharp and Li 1987a). Independent, single mutational events at this position will transiently result in a nonserine coding intermediate. This is not very likely for a highly conserved protein like ubiquitin. However, simultaneous double mutations are very rare events in nature and we proposed for sponge polyubiquitin gene a different explanation for the "apparent" double mutation at Ser codon 65 (Müller et al. 1994). The ancestral diubiquitin gene encoded Thr at position 65 and the corresponding codon was ACC. Thr and Ser are similar aa's. Single independent mutational events in both tandem repeats created Ser codons: ACC was changed to TCC (found in GCUBI1_5) or to AGC (found in GCUBI1_2, −4 and −6). Later mutation in the TCC codon resulted in the TCT codon (found in GCUBI1_1 and −3). Obviously, this Thr to Ser change was not silent. Not only was it permissive, it was evolutionarily favorable and was therefore selected by nature because it created a "more functional" ubiquitin (Müller et al. 1994). Ancestral genes encoding Thr_{65} disappeared from metazoan organisms and the remnants of their existence can be seen in the unusual usage of synonymous Ser codons.

4.1.2.2
Homology Comparison of the Repeated Genes

According to the "double mutation" which did not lead to an exchange of Ser_{65} and the respective other mutated sites in GCUBI1_1 to −6, the six repeats can be subdivided into group I (GCUBI1_1, −3 and −5) and group II (GCUBI1_2, −4 and 6; Müller et al. 1994). The average difference between group I and group II repeats is 22. Within the groups, repeats have considerably fewer substitutions and have been ranked according to decreasing homology: GCUBI1_1, −3 and −5 in group I, and GCUBI1_2, −4 and −6 in group II. Two repeats belonging to two groups are always linked together (GCUBI1_1 and −2; GCUBI1_2 and −4; GCUBI1_5 and −6). From this finding it is evident that the recent polyubiquitin gene (hexamer) was created by consecutive duplications of the ancestral dimeric gene. An evolution of the polyubiquitin gene in *G. cydonium* can be formulated as outlined in Fig. 2 (Müller et al. 1994). In the same figure, changes from Thr to Ser are also shown. These changes had already happened in the tandem genes (Y and Z), ancestral to group I and group II genes. Duplication of the diubiquitin gene resulted in the tetramer with the gene arrangement GCUBI1_1, −2, −5 and −6. The most recent event

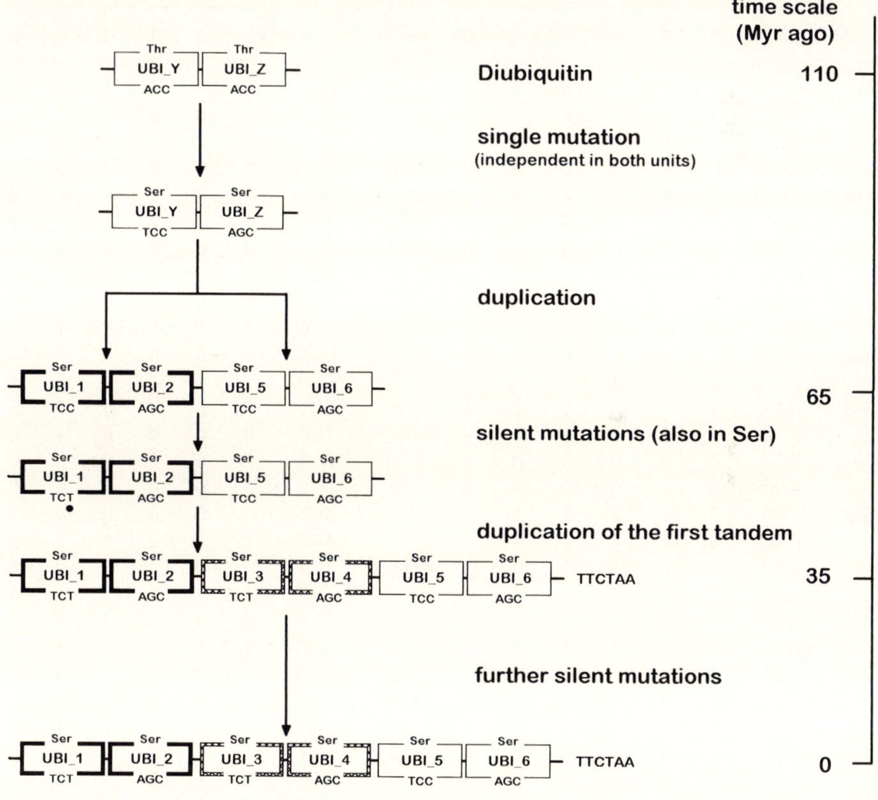

Fig. 2. Proposed evolution of *G. cydonium* polyubiquitin gene comprising the six repeats GCUBI1_1 to GCUBI1_6 (abbreviated UBI-1 to UBI-6). Changes from Thr to Ser in the proposed UBI_Y and UBI_Z diubiquitin are shown. Time scale on the *right* indicates the occurrence of each step of duplication

was duplication of GCUBI1_1 and −GCUBI1_2 tandem repeats in the tetramer, leading to the hexamer with the gene arrangement GCUBI1_1, −2, −3, −4, −5 and −6. In the recent sponge polyubiquitin gene, repeats 1 and 2 are the oldest, 5 and 6 are intermediate and 3 and 4 are the most recent acquisition (Fig. 2).

4.1.2.3
Codon Usage in the *G. cydonium* Polyubiquitin Gene

Codon usage in the *G. cydonium* polyubiquitin gene is listed in Table 2. It is obvious that the sponge polyubiquitin gene shows nonrandom usage of syn-

Table 2. Codon usage in *G. cydonium* polyubiquitin gene (GCUBI1_1 to GCUBI1_6)

| | | GCUBI1 | | | | | | | | GCUBI1 | | | | | |
		1	2	3	4	5	6			1	2	3	4	5	6
Ala	GCT	0	3	2	3	1	2		CTC	4	5	5	3	4	5
	GCC	3	0	1	0	2	1		CTA	1	0	0	0	1	2
	GCA	0	0	0	0	0	0		CTG	2	2	2	2	2	1
	GCG	0	0	0	0	0	0	Lys	AAA	0	0	0	0	0	1
Arg	CGT	3	2	3	2	3	3		AAG	7	7	7	7	7	6
	CGC	1	2	1	2	1	1	MET	ATG	1	1	1	1	1	1
	CGA	0	0	0	0	0	0	Phe	TTT	0	0	0	0	1	1
	CGG	0	0	0	0	0	0		TTC	2	2	2	2	1	2
	AGA	0	0	0	0	0	0	Pro	CCT	0	0	0	0	0	0
	AGG	0	0	0	0	0	0		CCC	1	1	1	1	1	1
Asn	AAT	0	0	0	0	0	1		CCA	0	0	0	0	0	0
	AAC	2	2	2	2	2	1		CCG	1	1	1	1	1	1
Asp	GAT	1	1	0	1	0	1	Ser	TCT	1	0	1	0	0	0
	GAC	4	4	5	4	5	4		TCC	0	0	0	0	1	0
Cys	TGT	0	0	0	0	0	0		TCA	0	0	0	0	0	0
	TGC	0	0	0	0	0	0		TCG	0	0	0	0	0	0
Gln	CAA	2	2	2	2	2	2		AGT	0	0	0	0	0	1
	CAG	4	4	4	4	4	4		AGC	2	3	2	3	2	2
Glu	GAA	2	2	3	2	3	2	Thr	ACT	0	0	0	0	0	0
	GAG	4	4	3	4	3	4		ACC	4	3	3	3	3	4
Gly	GGT	3	2	3	2	3	1		ACA	3	3	3	2	3	2
	GGC	1	3	2	3	0	2		ACG	0	1	1	2	1	1
	GGA	1	1	1	1	2	1	Trp	TGG	0	0	0	0	0	0
	GGG	1	0	0	0	1	2	Tyr	TAT	0	0	0	0	0	0
His	CAT	0	1	0	1	1	1		TAC	1	1	1	1	1	1
	CAC	1	0	1	0	0	0	Val	GTT	0	1	0	1	0	0
Ile	ATT	0	1	0	2	0	1		GTC	2	1	2	3	2	2
	ATC	7	6	7	5	7	6		GTA	1	0	1	0	0	0
	ATA	0	0	0	0	0	0		GTG	1	2	1	1	2	2
Leu	TTA	0	0	0	0	0	0	−	TAA	0	0	0	0	0	0
	TTG	0	0	0	0	0	0		TGA	0	0	0	0	0	0
	CTT	2	2	2	3	2	1		TAG	0	0	0	0	0	0

onymous codons. A biased pattern of codon usage is characteristic of highly expressed genes (Ikemura 1985). These genes have developed an optimal codon usage specific for a given organism and adapted to the amount of corresponding tRNA isoacceptors. In the *G. cydonium* polyubiquitin gene, the aa's Asp, Gln, Glu, Gly, His, Phe and Val have used all possible codons. However, among four possible codons for Ala only two are used; for Ile only one (out of three) is used with high frequency; and the same is true for Lys (one out of two) and some other aa's (Müller et al. 1994). In highly expressed genes, where synonymous codons are not used equally, the rate of silent substitutions is reduced (Sharp and Li 1987c). The nonrandom usage of synonymous codons in the sponge polyubiquitin gene might have had some influence on the rate of accumulation of silent substitutions.

4.1.2.4
Time Scale for Evolution of the *G. cydonium* Polyubiquitin Gene

A typical rate of the accumulation of silent changes was established for higher mammals (Li et al. 1987) and is, on average, 1% per site per five million years (Myr). However, huge differences were found in different species, even among mammals: for example, rodents have 4 to 8 times higher rates of the accumulation of silent mutations than primates, especially humans. Many parameters influence mutational rates i.e. generation time, size of the population, efficiency of the repair system, "life style" etc. Furthermore, the rate of silent substitutions was probably twice as high only 70 Myr ago (Sharp and Li 1987a). There is not enough sequence data from the lower Metazoa, nor reliable time distances between different groups, to establish their rate of accumulation of silent substitutions. Therefore, a rate of 1% per site per 5 Myr was accepted for sponge genes (Müller et al. 1994). According to the calculation proposed by Li et al. (1987), ubiquitin genes have about 50 silent (synonymous) sites per repeat (Sharp and Li 1987a) and sponge ubiquitin has 51.8 (Müller et al. 1994). Altogether, 22 changes were found between GCUBI1_1 and GCUBI1_2, the two oldest repeats in the sponge polyubiquitin gene. In other words, each gene separately accumulated 11 independent mutations (21.2% of changes). With a rate of 1% per site per 5 Myr, the time needed to accumulate this number of silent changes is about 106 Myr (Müller et al. 1994). Therefore, the dimeric gene must have existed at least 100 Myr ago. Duplication of the dimer, resulting in the tetramer, occured some 60–65 Myr ago (on average 12.55% of changes exist between repeats 1 and 5, or 2 and 6) and the youngest repeat (GCUBI1_3 and −4) was formed from GCUBI1_1 and −2 after the second duplication of this tandem, some 30–35 Myr ago (on average 6.5% changes from repeats 1 and 2; Müller et al. 1994). Time scale, indicating the occurrence of each step of duplication is shown on the right of Fig. 2.

At present, it is not possible to discuss how the calculated time scale reflects the real time of evolution of the sponge polyubiquitin gene. Many parameters, as already mentioned, can considerably influence the accumulation of the silent changes. In addition, rate of silent substitution is reduced in genes with biased codon usage (Sharp and Li 1987c) and concerted evolution of ubiquitin genes can further complicate the analysis of mutational events. There is some evidence in *G. cydonium* polyubiquitin gene for the homogenization of three central repeats (GCUBI1_2, −3 and −4) which share very homologous 5′-ends (Müller et al. 1994). However, unique silent substitutions found in most distant repeats (GCUBI1_1 and GCUBI1_6) and not shared with any other repeat, also permitted calculation of the time scale (Müller et al. 1994) and are fully in agreement with the time proposed for the creation of the sponge tetramer (60–65 Myr ago) or hexamer (30–35 Myr ago). The dimeric ancestral gene must have existed much more than 100 Myr ago, since class III polyubiquitin genes in all analysed eukaryotes have at least one tandem repeat of ubiquitin genes. A relatively small difference between the two most ancient repeats (21%) is obviously the consequence of ancient processes of gene homogenization, as well as many back and forth mutations.

Fossil species belonging to the family of Geodiidae (Gray) 1867, range as far back as the Carboniferous period (280–225 Myr ago) and that of the genus *Geodia* (Lamarck) 1815, have Cretaceous (145–75 Myr ago) to recent range (De Laubenfels 1955). If the time scale for the creation of polyubiquitin in *G. cydonium* is (close to) correct, than our finding implies that we are witnessing the process of the evolution of the diubiquitin gene into the recent polyubiquitin gene (hexamer) in the genus *Geodia*.

4.1.3
Is There One Additional Polyubiquitin Gene in Geodia cydonium?

Recently, a cDNA clone also encoding polyubiquitin was isolated from a library prepared from another specimen of *G. cydonium*, collected from the same area (Rovinj, northern Adriatic Sea, Croatia). The aim was to look for intraspecies variations in the polyubiquitin gene. To our surprise this cDNA clone showed a considerably different nt sequence; termed GCUBI2. The size of the clone (1.6 kb), number of repeats (six) and aa sequence of the encoded ubiquitin are identical as already reported for GCUBI1 (Pfeifer et al. 1993a; Müller et al. 1994). Immediate 5′- and 3′-untranslated sequences are also highly homologous to those in GCUBI1 and the extra aa in the last repeat is Phe. So far, only gene repeats 1, 2, 5 and 6 were fully sequenced and analysed (Fig. 3A,B). They display on average 13 silent changes (12 to 15) from the corresponding repeats in GCUBI1 and similar values were obtained also with noncorresponding repeats (i.e. repeat 1 of GCUBI2 [GCUBI2_1]:GCUBI1_1 = 12 differences; GCUBI2_1:GCUBI1_2 = 13 differences; GCUBI2_1:GCUBI1_3 = 11 differences). Furthermore, repeats 1 and 2 of GCUBI2 are very similar to each other (only six silent changes) and use the same codon (TCT) for Ser_{65}; this is a completely different situation from that found between GCUBI1_1 and GCUBI1_2. It is evident that in this polyubiquitin gene, the process of concerted evolution homogenized several genes inside the polygene (partial sequence of the repeat 4 is known, and Ser_{65} is also encoded by TCT). The last repeat in the polygene (GCUBI2_6) uses AGC for Ser_{65} (like in GCUBI1_6) and does not seem to be included in the homogenization process, at least not at the 3'-end. However, it still shows 12 silent changes when compared with GCUBI1_6. These findings raise the question about the number of polyubiquitin genes in *G. cydonium*. Screening of the first cDNA library identified only one (expressed) gene. All five analysed polyubiquitin cDNAs had identical sequences (named GCUBI1). If there is really only one polyubiquitin gene in *G. cydonium*, then two individual sponges, living in the same area, contain unusually different gene(s). Differences in the coding sequence between two (identical?) polyubiquitin genes in *G. cydonium* are over 5%, much more than the average differences found between two related species. Only extreme sexual isolation, in connection with the hypermutability of the sponge genome, can result in such intra species differences as were found in two *G. cydonium* polyubiquitin genes, which, in addition, are known to be the most conserved genes in eukaryotes. Therefore, it is very likely that at least two nonallelic polyubiquitin genes, with a different way of evolving and a different way of expression, are present in the genome of *G. cydonium*.

Multiple sequence alignment of ubiquitin repeats from two *G. cydonium* polyubiquitin cDNA clones and the dendrogram of phylogenetic relations between repeats, produced by the Clustal W program (Higgins and Sharp 1988; Thompson et al. 1994), are shown in Fig. 3.

4.2
Diubiquitin Gene from *Suberites domuncula*

The marine sponge *Suberites domuncula* is only distantly related to *G. cydonium*, although they both belong to Demospongiae. A cDNA library of *S. domuncula* was recently prepared in λZAP expression vector and screened for polyubiquitin encoding clones using polyubiquitin cDNA from *G. cydonium*. Only one cDNA clone was identified, and it encodes a diubiquitin gene (EMBL accession number Y12082). Search for longer clones is still in progress. Both repeats in the dimeric gene encode ubiquitin with an aa sequence identical to *G. cydonium* ubiquitin (Table 1), including the extra aa in the second (last) repeat (Phe). The two repeats differ from each other at 22 silent positions, the same degree as was found i.e. between GCUBI1_1 and GCUBI1_2 repeats. However, the differences between *S. domuncula* and *G. cydonium* ubiquitin repeats are very high (often over 40 changes), although all changes are silent. Moreover, in *S. domuncula* genes, all three Ser in ubiquitin are encoded by TCN codons (TCT for Ser_{20}, TCC for Ser_{57} and TCC for Ser_{65}) while in *G. cydonium*, Ser_{20} and Ser_{57} are encoded by AGPy in all ten analysed repeats, and only Ser_{65} is, as already discussed, encoded by serine codons from both families (TCN and AGPy) (Table 3, Sect. 4.4.). Ser codons AGT and AGC are used very rarely in ubiquitin genes. These codons were never used for Ser_{57} or Ser_{65} in "consensus" ubiquitin sequences of animal genes (Müller et al. 1994).

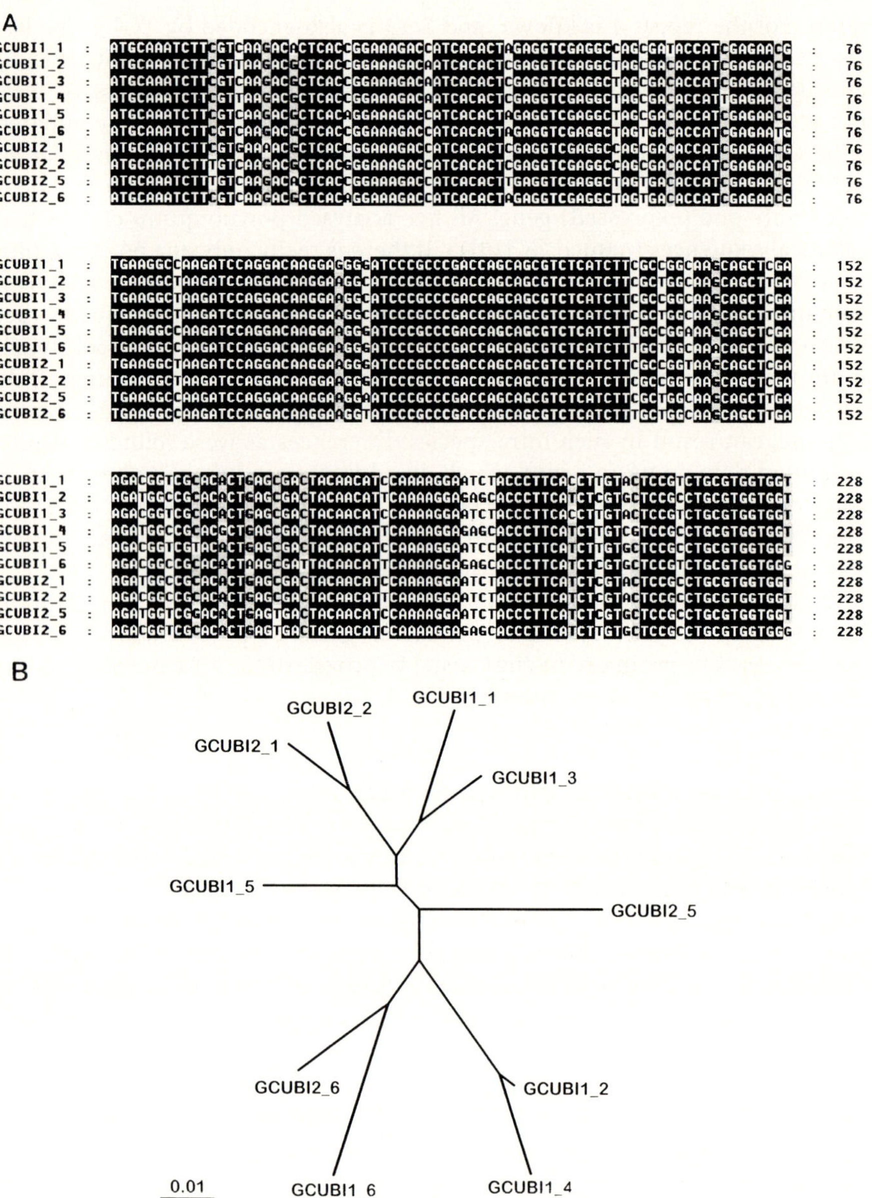

Fig. 3. A Multiple sequence alignment of ubiquitin gene repeats from two *G. cydonium* polyubiquitin genes from the two specimens under investigation; GCUBI1 (repeats GCUBI1_1 to GCUBI1_6) and GCUBI2 (GCUBI2_1 to GCUBI2_6). **B** Unrooted phylogenetic tree of the alignment. The analysis was performed by neighbour-joining and distance matrix determinations. Scale *bar* represents 1 change per 100 nt

Table 3. Usage of codons for Ser and Leu in sponge polyubiquitin genes

	Geodia[a]	_Suberites_[a]	_Sycon_[a]
Ser20	AGC/t	TCT	AGC/T
Ser57	AGC/t[b]	TCC	AGC/t
Ser65	TCT/c; AGC	TCC	TCC/t
Leu8	CTC	CTC	CTC/A,G
Leu15	CTT/C,A	CTT	TTG
Leu43	CTC	CTC/T	CTC/t
Leu50	CTC/t	CTC/T	CTC
Leu56	CTG/a	CTG/T	TTG/a
Leu67	CTT	CTT	CTT/c
Leu69	CTC/T	TTG	TTG
Leu71	CTC	CTC	T/CTG
Leu73	CTG	CTC	CTA/c

[a] Ten, two and seven ubiquitin gene repeats from _G. cydonium_, _S. domuncula_ and _S. raphanus_, respectively, were included in the analysis. [b] Lower case letters indicate less frequent usage of the alternative codon.

However, Ser_{20} is encoded by AGPy (apart from in _G. cydonium_) in all analysed vertebrates, but not in invertebrates and protozoa (Müller et al. 1994). Furthermore, Leu_{69} is in all _G. cydonium_ ubiquitin repeats encoded by CTC or CTT codons and in all analysed animal "consensus" sequences Leu_{69} is also encoded by CTN codons. In _S. domuncula_, both ubiquitin gene repeats use TTG for Leu_{69} (Table 3). Although changes of codons for Leu_{69} can be achieved by two subsequent silent substitutions, differences between the two sponges in codon usage for serines involve (very rare) simultaneous double substitutions and raise again the question about the origin of serines in ubiquitin. The easiest explanation is that the ancestral gene, common to both Demospongiae, encoded threonines instead of serines. Thr codons (ACN) can change to Ser codons (TCN or AGPy) by a single mutation.

Differences between ubiquitin genes from _G. cydonium_ and _S. domuncula_ are in the same range as differences between sponges and higher animal ubiquitin genes. This indicates that Demospongiae are an extremely old class of metazoan organisms and that _G. cydonium_ and _S. domuncula_ shared a common ancestor several hundred Myr ago.

4.3
Polyubiquitin Gene from _Sycon raphanus_

Sycon raphanus, a very small marine sponge, with a skeleton built of calcium carbonate, belongs to the class Calcispongiae. A cDNA library of _S. raphanus_, prepared in λZAP expression vector, was also screened for polyubiquitin encoding clones using polyubiquitin cDNA from _G. cydonium_. Several cDNA clones were identified and they all contained inserts of the same size (1.9kb).

Nucleotide sequences of the 5′- and 3′-ends of cDNAs are very similar but not identical, especially not at the 3′-end, indicating that several similar polyubiquitin genes are present in the genome of *S. raphanus*. Only one cDNA clone was fully analysed and this clone encodes a polyubiquitin precursor protein with seven ubiquitin units (EMBL accession number Y10526). All seven repeats in the polygene encode the same protein. Ubiquitin in *S. raphanus* differs at two positions from *G. cydonium* or *S. domuncula* ubiquitin (Table 1) and has at the 3′-end of the last repeat a different, extra aa (Cys). As in humans and all other higher animals, aa at position 19 in *S. raphanus* ubiquitin is Pro (Ala in *G. cydonium* and *S. domuncula*). However, the aa at position 24 is Asp, instead of Glu, found in all metazoan animals including two Demospongiae. Asp at position 24 is typical for yeast and plants (Table 1).

As in *G. cydonium*, polyubiquitin gene in *S. raphanus* also evolved by the duplication of the early dimer (i.e. repeats 1 and 2). Repeats 1, 3, and 5 cluster together, and the second group contains repeats 2, 4 and 6 (Fig. 4). The origin of repeat 7 at the 3′-terminus of *S. raphanus* polyubiquitin gene is uncertain, because it displays equal numbers of silent changes with repeats from both groups.

Analysis of the codon usage for Ser at positions 20, 57 and 65 shows (unexpected) similarities between *G. cydonium* and *S. raphanus* (Table 3, Sect. 4.2.). As in *G. cydonium*, Ser_{20} and Ser_{57} in *S. raphanus* ubiquitin are encoded by

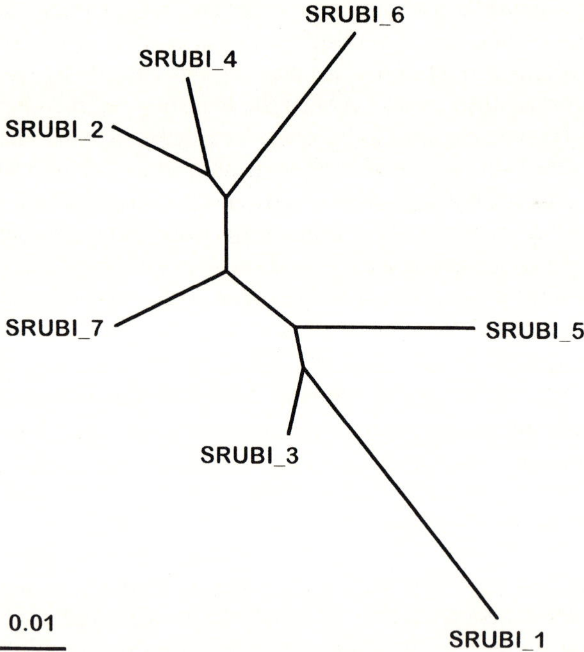

Fig. 4. Unroooted phylogenetic tree of seven ubiquitin gene repeats from *S. raphanus*; SRUBI_1 to SRUBI_7. Scale *bar* represents 1 change per 100 nt

AGPy codons. However, for Ser_{65} only TCC and TCT codons are used in all seven *S. raphanus* repeats. All nine leucines in *G. cydonium* ubiquitin (and eight in *S. domuncula*) are encoded by CTN codons. In *S. raphanus*, CTN and TTPu codons for leucine are used equally (including Leu_{69}). As expected, differences in the nt sequence between *S. raphanus* ubiquitin gene repeats and repeats from *G. cydonium* or *S. domuncula* are very high, but not higher than differences found between repeats from two Demospongiae.

4.4
Phylogenetic Relationships of Sponge Ubiquitin Genes

Polyubiquitin genes from the three sponges analysed differ to a very large extent, reaching almost saturation in silent changes, which indicates that many unregistered mutations must have happened since their separation from the common ancestor; calculations of the separation times are therefore not possible. Observed differences in sponge polyubiquitin genes, both in primary structure and number of repeats, are a reflection of the very old separation of sponges belonging to two classes of the subphylum Cellularia, or even the same class, Demospongiae. A dimeric diubiquitin gene was most probably the common ancestral gene in all sponge lineages and polyubiquitin genes in *G. cydonium* and *S. raphanus* were created long after their branching off from the common ancestor. Before the duplications of the dimeric ubiquitin genes, changes in the encoded aa's and in codon usage were established and Thr(s) in the ancestral ubiquitin were converted into Ser(s). Two protists, *G. lamblia* (Krebber et al. 1994) and *T. vaginalis* (Keeling and Doolittle 1995) still encode Thr at position 20 in ubiquitin genes. Moreover, *G. lamblia* is the only known organism to contain only a monoubiquitin gene and *T. vaginalis* is a member of one of the earliest diverging eukaryotic lineages (Cavalier-Smith 1993).

Differences in codon usage for Ser and Leu in sponge polyubiquitin genes are listed in Table 3. Ser is encoded by six codons belonging to two codon families and alternative usage of codons from two codon families can be achieved only by simultaneous double mutation. Leu (as well as Arg) is also encoded by six codons; however, changes from one codon to another can proceed without the change in the encoded aa. For example, change of Leu/TTA to Leu/CTG, which looks like a double mutation, can proceed via a Lue/TTG or Leu/CTA intermediate. A second silent change will finaly create Leu/CTG. Still, an alternative codon usage in ubiquitin repeats, like TTG and CTC for Leu at the same position in the protein, indicates that the change happened before the duplications of the ancestral gene started.

Ubiquitin gene repeats from three sponges were aligned and analysed using the CLUSTAL W program (Thompson et al. 1994) and the unrooted phylogenetic tree produced by this analysis is shown in Fig. 5. All gene repeats from one organism cluster together, as was generally found with ubiquitin genes. Ubiquitin gene sequences from any particular species always form a coherent group and this is a very strong indication for the concerted evolution of

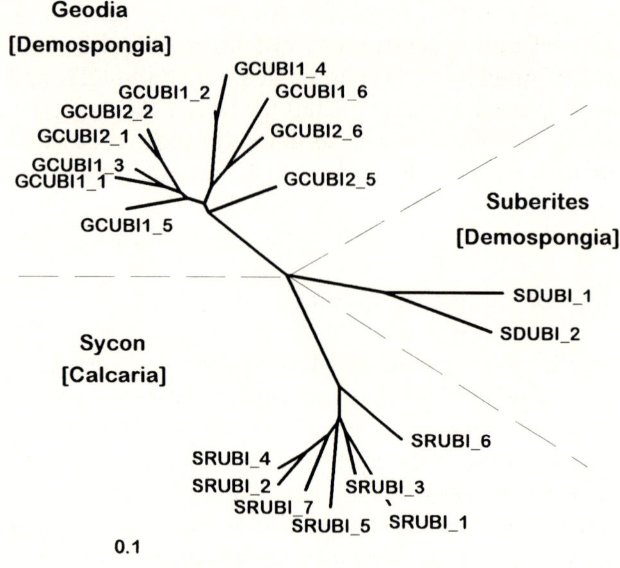

Fig. 5. Unrooted phylogenetic tree of ubiquitin gene repeats from the three marine sponges *G. cydonium* (GCUBI1 and GCUBI2); *S. domuncula* (SDUBI) and *S. raphanus* (SRUBI), studied here. Scale *bar* represents 1 change per 100 nt

ubiquitin genes. As expected, two Demospongiae, *G. cydonium* and *S. domuncula* do not cluster together, although they possess identical ubiquitins. However, differences in their ubiquitin gene repeats are very high, as high as differences found between *S. raphanus* and the two Demospongiae.

5
Phylogenetic Tree of Metazoa Based on Polyubiquitin Genes

In an effort to construct a phylogenetic tree of ubiquitin genes, polyubiquitin genes from 14 animal species, including three sponges and two protozoans, were compared. Polyubiquitin genes contained a minimum of one repeat (*G. lamblia*) and a maximum of 11 repeats (*C. elegans*) and only one cluster of ubiquitin genes was analysed in each species. Repeats from the same polyubiquitin gene were first aligned and compared and a "consensus", species-specific monoubiquitin sequence was deduced for each species (Müller et al. 1994). In the case of sponges, sequences of all ubiquitin repeats were taken into the analysis. Note that ubiquitins are identical from *Drosophila* to man and their "consensus" sequences reflect only the differences in codon usage. An unrooted phylogenetic tree is shown in Fig. 6. The phylogenetic tree of ubiquitin genes produced by this alignment is in good agreement with the established phylogenetic relationships among analysed animals. All verte-

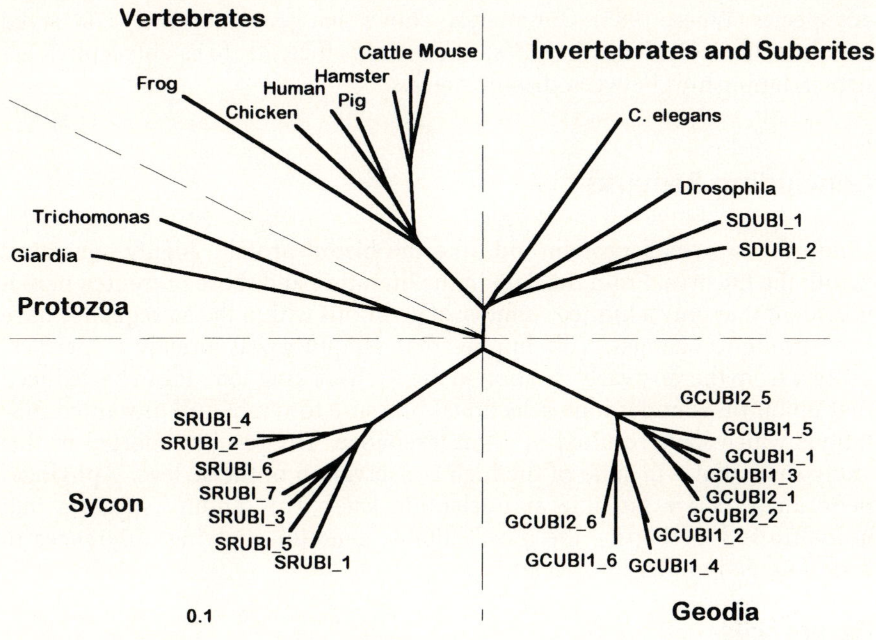

Fig. 6. Phylogenetic tree of animals based on the ubiquitin genes. Polyubiquitin genes with following accession numbers were used for this analysis: U27577 (protozoa; *Trychomonas vaginalis*), X70050 (protozoa; *Giardia lamblia*), M23433 (nematode; *Caenorhabditis elegans*), M22428 (insect; *Drosophila melanogaster*), M11512 (frog; *Xenopu laevis*), X58195 (chicken), M18159 (pig), Z18245 (cattle), X60390 (hamster), X51703 (mouse) and M26880 (human). In addition the sponge sequences are included. Scale *bar* represents 10 changes per 100 nt

brates are clustered together. *G. cydonium* and *S. raphanus* ubiquitin genes make a separate group. However, *S. domuncula* is clustered together with invertebrates, most probably as a consequence of the higher similarity in codon usage, especially for Ser. Ubiquitins in *G. cydonium, S. domuncula* and *C. elegans* (Nematoda) are identical proteins (Table 1), although gene sequences between i.e. *G. cydonium* and *C. elegans* differ at 38–44 positions (Müller et al. 1994), the same difference as found between *G. cydonium* genes and mammalian genes (39–45). In the previous analysis, when only *G. cydonium* ubiquitin gene "consensuses" were compared with "consensuses" of 11 different animals (Müller et al. 1994), sponge genes clustered together with invertebrate genes and showed the earliest branching off from the common ancestral ubiquitin gene.

From the differences found in ubiquitin "consensus" sequences from distant species, it is not possible to calculate times when these genes could have diverged. Many unregistered mutations (back and forth) in ubiquitin genes must have occurred during evolution. However, present differences in ubiquitin genes still reflect very well the phylogenetic relations between different species (Woese 1983). Our analysis shows that genes extremely conserved in nature, like polyubiquitin genes, can be used efficiently to establish phylogenetic relationships between distant species.

6
Concluding Remarks

Ubiquitin is a small protein and, like the histon protein, highly conserved within the Eucarya. From the data in the literature and those presented here it is evident that only a limited number of positions within the aa sequence have been prone to changes. This implies that ubiquitin was already a "perfect" protein from the very early creation in the Archaea kingdom. It can be deduced that ubiquitin is under high selectional pressure to remove all unwanted mutations which have resulted in an aa exchange. The results reported in this study show that – in spite of the high conservation of the aa level – phylogenetic analyses are possible at nucleotide level. This finding suggests that ubiquitin is – perhaps – the most suitable gene for studying differences in codon usage in different species.

References

Bachmair A, Finley D, Varshavsky A (1986) In vitro half-time of the protein is a function of its amino-terminal residue. Science 234:179–186

Baker RT, Board PG (1987) The human ubiquitin gene family: structure of a gene and pseudogenes from the UB B subfamily. Nucleic Acids Res 15:443–463

Baker RT, Tobias JW, Varshavsky A (1992) Ubiquitin specific proteases of *Saccharomyces cerevisiae*. J Biol Chem 267:23364–23375

Biesalski HK, Doepner G, Tzimas G, Gamulin V, Schröder HC, Batel R, Nau H, Müller WEG (1992) Modulation of myb gene expression in sponges by retinoic acid. Oncogene 7:1765–1774

Bond V, Schlesinger MJ (1986) The chicken ubiquitin gene contains a heat-shock promoter and expresses an unstable mRNA in heat-shocked cells. Mol Cell Biol 6:4602–4610

Callis J, Carpenter T, Sun C-W, Vierstra RD (1995) Structure and evolution of genes encoding polyubiquitin and ubiquitin-like proteins in *Arabidopsis thaliana* ecotype Columbia. Genetics 139:921–939

Cavalier-Smith T (1993) Kingdom Protozoa and its 18 phyla. Microbiol Rev 57:953–994

De Laubenfels MW (1955) Archaeocyta and porifera. In: Moore RC (ed) Treatise on invertebrate paleontology, part E. Geological Society of American University of Kansas Press, Kansas, pp 22–122

Doherty F, Mayer RJ (1992) Intracellular protein degradation. IRL Press, Oxford University Press, New York

Dworkin-Rastl E, Shrutkowski A, Dworkin MB (1984) Multiple ubiquitin mRNAs during *Xenopus laevis* development contain tandem repeats of the 76 amino acid coding sequence. Cell 39:321–325

Einspanier R, Sharma HS, Scheit KH (1987) An mRNA encoding polyubiquitin in porcine corpus luteum: identification by cDNA cloning and sequencing. DNA 6:395–400

Eytan E, Ganoth A, Armon Y, Hershko A (1989) ATP-dependent incorporation of 20S protease into the 26S complex that degrades proteins conjugated to ubiquitin. Proc Natl Acad Sci USA 86:7751–7755

Finley D, Chau V (1991) Ubiquitination. Annu Rev Cell Biol 7:25–69

Finley D, Varshavsky A (1985) The ubiquitin system: functions and mechanisms. Trends Biochem Sci 10:343–346

Finley D, Özkaynak E, Varshavsky A (1987) The yeast polyubiquitin gene is essential for resistance to high temperatures, starvation and other stresses. Cell 48:1035–1046

Finley D, Bartel B, Varshavsky A (1989) The tails of ubiquitin precursors are ribosomal proteins whose fusion to ubiquitin facilitates ribosome biogenesis. Nature 338:394–401

Gamulin V, Rinkevich B, Schäcke H, Kruse M, Müller IM, Müller WEG (1994) Cell adhesion receptors and nuclear receptors are highly conserved from the lowest Metazoa (marine sponges) to vertebrates. Biol Chem Hoppe Seyler 375:583–588

Glotzer M, Murray AW, Kirschner MW (1991) Cyclin is degraded by the ubiquitin pathway. Nature 349:132–138

Goebl MG, Yochem J, Jentsch S, McGrath JP, Varshavsky A, Byers B (1988) The yeast cell cycle gene CDC34 encodes a ubiquitin-conjugating enzyme. Science 241:1331–1335

Goldstein G, Scheid M, Hammerling U, Boyse EA, Schlesinger DH, Niall HD (1975) Isolation of a polypeptide that has lymphocyte-differentiating properties and is probably represented universally in living cells. Proc Natl Acad Sci USA 72:11–15

Graham RW, Jones D, Candido PM (1989) UbiA, the major polyubiquitin locus in *Caenorhabditis elegans*, has unusual structural features and is constitutively expressed. Mol Cell Biol 9:268–277

Gramzow M, Schröder HC, Fritsche U, Kurelec B, Robitzki A, Zimmermann H, Friese K, Kreuter MH, Müller WEG (1989) Role of phospholipase A2 in the stimulation of sponge cell proliferation by the homologous lectin. Cell 59:939–948

Gropper K, Brandt RA, Elias S, Bearer CF, Mayer A, Schwartz AL, Ciechanover A (1991) The ubiquitin activating enzyme, E1, is required for stress-induced lysosomal degradation of cellular proteins. J Biol Chem 266:3602–3610

Henkart P, Humphreys S, Humphreys T (1973) Characterization of sponge aggregation factor. A unique proteoglycan complex. Biochemistry 12:3045–3050

Hershko A, Ciechanover A (1992) The ubiquitin system for protein degradation. Annu Rev Biochem 61:761–807

Higgins DG, Sharp PM (1988) CLUSTAL: a package for performing multiple sequence alignment on a microcomputer. Gene 73:237–244

Ikemura T (1985) Codon usage and tRNA content in unicellular and multicellular organisms. Mol Biol Evol 2:13–34

Jentsch S (1992) Ubiquitin dependent protein degradation: a cellular perspective. Trends Cell Biol 2:98–103

Jentsch S, McGrath JP, Varshavsky A (1987) The DNA repair gene RAD6 encodes a ubiquitin-conjugating enzyme. Nature 329:131–134

Jentsch S, Seufert W, Hauser HP (1991) Genetic analysis of the ubiquitin system. Biochim Biophys Acta 1089:127–139

Keeling PJ, Doolittle WF (1995) Concerted evolution in protists: recent homogenization of a polyubiquitin gene in *Trichomonas vaginalis*. J Mol Evol 41:556–562

Kozak M (1984) Compilation analysis of sequences upstream from the translation start site in eukaryotic mRNA. Nucleic Acids Res 12:857–872

Krebber H, Wöstmann C, Bakker-Grunwald T (1994) Evidence for the existence of a single ubiquitin gene in *Giardia lamblia*. FEBS Lett 343:234–236

Laszlo L, Doherty FJ, Osborn NU, Mayer RJ (1990) Ubiquitinated protein conjugates are specifically enriched in the lysosomal system of fibroblasts. FEBS Lett 261:365–368

Li W-H, Tanimura M, Sharp PM (1987) An evaluation of the molecular clock hypothesis using mammalian DNA sequences. J Mol Evol 25:330–342

Mayer AN, Wilkinson KD (1989) Detection, resolution and nomenclature of multiple ubiquitin carboxyl-terminal esterases from bovine calf thymus. Biochemistry 28:166–172

Mayer RJ, Arnold J, Laszlo L, Landon M, Lowe J (1991) Ubiquitin in health and disease. Biochim Biophys Acta 1089:141–157

Müller WEG (1995) Molecular phylogeny of Metazoa (animals): monophyletic origin. Naturwissenschaften 82:321–329

Müller WEG, Zahn RK (1973) Purification and characterization of a species-specific aggregation factor in sponges. Exp Cell Res 80:95–104

Müller WEG, Diehl-Seifert B, Gramzow M, Friese U, Renneisen K, Schröder HC (1988) Interrelation between extracellular adhesion proteins and extracellular matrix in reaggregation of dissociated sponge cells. Int Rev Cytol 111:211–229

Müller WEG, Ugarkovic D, Gamulin V, Weiler BE, Schröder HC (1990) Intracellular signal transduction pathways in sponges. Electron Microsc Rev 3:97–114

Müller WEG, Schröder HC, Müller IM, Gamulin V (1994) Phylogenetic relationship of ubiquitin repeats of the polyubiquitin gene from the marine sponge *Geodia cydonium*. J Mol Evol 39:369–377

Müller WEG, Müller IM, Schröder HC, Gamulin V (1995a) On the monophyletic evolution of the Metazoa. Braz J Med Biol Res 27:2083–2097

Müller WEG, Müller IM, Rinkevich B, Gamulin V (1995b) Molecular evolution: evidence for the monophyletic origin of multicellular animals. Naturwissenschaften 82:36–38

Nagylaki T (1984) The evolution of multigene families under intrachromosomal gene conversion. Genetics 106:529–548

Neves A, Guerreiro P, Rodrigues-Pousada C (1990) Striking changes in polyubiquitin genes of *Tetrahymena pyriformis*. Nucleic Acids Res 18:656

Nickel BE, Allis CD, Davie JR (1989) Ubiquitinated histone H2B is preferentially located in transcriptionally active chromatin. Biochemistry 28:958–963

Ohmachi T, Giorda R, Shaw DR, Ennis HL (1989) Molecular organization of developmentally regulated *Dictyostelium discoideum* ubiquitin cDNAs. Biochemistry 28:5226–5231

Ohta T (1980) Evolution and variation of mutligene families. Springer, Berlin Heidelberg New York

Ohta, T (1984) Some models of gene conversion for treating the evolution of multigene families. Genetics 106:527–528

Ohta T, Dover GA (1983) Population genetics of multigene families that are dispersed into two or more chromosomes. Proc Natl Acad Sci USA 80:4079–4083

Özkaynak E, Finley D, Varshavsky A (1984) The yeast ubiquitin gene: head-to-tail repeats encoding a polyubiquitin precursor protein. Nature 312:663–666

Özkaynak E, Finley D, Varshavsky A (1987) The yeast ubiquitin genes: a family of natural gene fusions. EMBO J 6:1429–1439

Pfeifer K, Frank W, Schröder HC, Gamulin V, Rinkevich B, Batel R, Müller IM, Müller WEG (1993a) Cloning of the polyubiquitin cDNA from the marine sponge *Geodia cydonium* and its preferential expression during reaggregation of cells. J Cell Sci 106:545–554

Pfeifer K, Haasemann M, Gamulin V, Bretting H, Fahrenholz F, Müller WEG (1993b) S-type lectins occur also in invertebrates: high conservation of the carbohydrate recognition domain in the lectin genes from the marine sponge *Geodia cydonium*. Glycobiology 6:179–184

Schäcke H, Schröder HC, Gamulin V, Rinkevich B, Müller IM, Müller WEG (1994) Molecular cloning of a receptor tyrosine kinase from the marine sponge *Geodia cydonium*: a new member of the receptor tyrosine kinase class II family in invertebrates. Mol Membr Biol 11:101–107

Schlesinger DH, Goldstein G (1975) Molecular conservation of 74 amino acid sequence of ubiquitin between cattle and man. Nature 255:423–424

Schwartz LM, Myer A, Kosz L, Engelstein M, Maier C (1990) Activation of polyubiquitin gene expression during developmentally programmed cell death. Neuron 5:411–419

Sharp PM, Li W-H (1987a) Molecular evolution of ubiquitin genes. Trends Ecol Evol 2:328–332

Sharp PM, Li W-H (1987b) Ubiquitin genes as a paradigm of concerted evolution of tandem repeats. J Mol Evol 25:58–64

Sharp PM, Li W-H (1987c) The codon adaptation index – a measure of directional synonymous codon usage bias, and its potential application. Nucleic Acids Res 15:1281–1295

Short JM, Fernandez J, Sorger JA, Huse WD (1988) Lambda ZAP: a bacteriophage lambda expression vector with in vivo excision properties. Nucleic Acids Res 7583–7600

Swindle J, Ajioka J, Eisen H, Sanwal B, Jacquemot C, Browder Z, Buck G (1988) The genomic organization and transcription of the ubiquitin genes of *Trypanosoma cruzi*. EMBO J 7:1121–1127

Thompson JD, Higgins DG, Gibson TJ (1994) CLUSTAL W: improving the sensitivity of progressive multiple sequence alignment through sequence weighting, position-specific gap penalties and weight matrix choice. Nucleic Acids Res 22:4673–4680

Vijay-Kumar S, Bugg CE, Cook WJ (1987a) Structure of ubiquitin refined at 1.8 Å resolution. J Mol Biol 194:531–544

Vijay-Kumar S, Bugg CE, Wilkinson KD, Vierstra RD, Hatfield P (1987b) Comparison of the three-dimensional structures of human, yeast and oat ubiquitin. J Biol Chem 262:6396–6399

Wenzel T, Baumeister W (1993) *Thermoplasma acidophilum* proteasomes degrade partially unfolded and ubiquitin-associated proteins. FEBS Lett 326:215–218

Wiborg O, Pedersen MS, Wind A, Berglund LE, Marcker KA, Vuust J (1985) The human ubiquitin multigene family: some genes contain multiple directly repeated ubiquitin coding sequences. EMBO J 4:755–759

Woese CR (1983) The primary lines of descent and universal ancestor. In: Bendall DS (ed) Evolution from molecules to man. Cambridge University Press, Cambridge, pp 209–233

Wofl S, Lottspeich F, Baumeister W (1993) Ubiquitin in the archaebacterium *Thermoplasma acidophilum*. FEBS Lett 326:42–44

Wöstmann C, Tannich E, Bakker-Grunwald T (1992) Ubiquitin of *Entamoeba histolytica* deviates in six amino acid residues from the consensus of all other known ubiquitins. FEBS Lett 308:54–58

Wray CG, DeSalle R (1994) Phylogenetic utility of ubiquitin DNA sequence from 3 marine protist lineages. Mol Mar Biol Biotechnol 3:13–22

Zarkower D, Stephenson P, Sheets M, Wickens M (1986) The AAUAAA sequence is required both for cleavage and for polyadenilation of Simian virus 40 pre-mRNA in vitro. Mol Cell Biol 6:2317–2323

Subject Index

Printing: Saladruck, Berlin
Binding: Buchbinderei Lüderitz & Bauer, Berlin